Outer Space in Washington DC

A tourist's guide

Peter Bassett F.R.A.S.

Astronomy Roadshow Publishing

Further paperback / signed / discounted copies and an e-book version with direct internet links can be purchased via…

www.outerspacebooks.com

The author took virtually all images. Credits to other images have been given wherever possible.

Proof read by C Beevis, info@authorcassclark.com

**A second book in this series is available –
Outer Space across the UK.
More are planned.**

Contents   Page

1   Introduction   3
2   Maps and Transport   5
3   Arlington Cemetery   10
4   Einstein Memorial   64
5   Werner von Braun Gravesite   68
6   US Naval Observatory   73
7   National Air & Space Museum   84
8   Dulles Air & Space Museum   93
9   Natural History Museum   96
10   NASA Headquarters   99
11   Capitol Building   101
12   NOAA Sites   104
13   Hubble Operations Center   107
14   Goddard Spaceflight Center   110
15   Dining Out in DC   114
16   Links for Reference   119
17   Other Books by the Author   120

The Metro is the key to exploring Washington DC.
Just follow the brown 'M' signs around the city.

Chapter 1   Introduction

Many people interested in Spaceflight / Astronomy love to include related places of interest while on vacation. After enduring many tours around the globe, I have never witnessed a city such as Washington DC with so many space sites in such a small area. This book illustrates them as well as a few others in the vicinity. It is kept straight to the point and never side-track.

Some chapters offer tips on how to get the most out of each site taking into account weather, walking distance, refreshments, security requirements etc. I have literally lost count of the number of times my wife Amanda and I have visited Washington DC. With every encounter, we try to discover something new and make notes to improve the experience of the next trip.

Having explored virtually every US city from Atlanta to Yuma, Washington DC is the finest place to visit regarding space-related sites. Hence the production of this book!

Moving around town by car is not advised. The traffic restrictions, parking, delays and various diversions simply turn the experience into a nightmare. Travelling around the city is best done either on foot, Metro Subway or by bus. Plan to stay at a Hotel that is within walking distance of a Metro station. We have used the Americana Hotel, close to Crystal City Metro station, next to the Pentagon. In 2023, it was sadly demolished. They offered a free bus service to and from the major airports. If you inadvertently drive past the hotel, you will automatically arrive in the parking lot of The Pentagon. Simply turn around and double-back. However, be advised not to get out and take a picture or you will be confronted by the Secret Service within seconds – (from harsh experience!).

You can purchase Metro tickets to any value via any Metro SmarTrip blue vending machine. Either buy for a single journey or top up your card and use it until it runs out. Each journey subtracts from your balance. Once it reaches zero and you try to exit, you will be instructed to top it up at another machine before letting you through the barrier. (N.B. You will not be arrested).

On Foot

Never underestimate the walking distance between sites. Unless you are an energetic youngster, I would recommend conserving your energy by using public transport. During my first solo visit on 30 March 1990, when I was 26, I managed to explore the city from Arlington Cemetery to Union Station in 16hrs, with hardly a break or using the Metro at all. I then drove on to West Virginia and slept in the car. (Now in my 50s there's no chance of me doing that anymore, as these days I would be lucky to get halfway round Arlington before needing to sit down!).

On that first evening in DC, trying to head for West Virginia, I did get a little lost somewhere in the North West district past Georgetown looking for Interstate 66. I pulled up and asked an innocent looking young woman wearing a short skirt for directions. She replied, *"Is that all you want honey? For $50 you can have anything you desire."* I drove on and found an old guy, (who wasn't wearing a short skirt …).

Every late March / early April, there is the National Cherry Blossom Festival. The city is pretty, out in full bloom, but does increase the hotel costs. www.nationalcherryblossomfestival.org

Chapter 2   Maps & Transport

Navigating around the city is very simple. Two main points of reference are the Washington Monument, and the Capitol Building. Both are clearly visible from the 12,000ft long National Mall, making it easy to orientate yourself with your tourist map. Do purchase a detailed map that marks all the buildings, monuments and museums. Obtain one in advance of your trip if possible, (you can find them on Amazon), so that you can plan and explore the City area by area. To see all the sites mentioned in this book plus the other 'must see' attractions, allow at least six days for a non-rushed experience.

Any City guide as shown above will enhance your visit. Ensure it includes a map of the Metro system. Some hotels freely provide them.

Search for the coupon books as shown left. They are often found at 'Subway' outlets, and rest areas along the Interstate. Discounts are given for hotels and food outlets.

Metro-Subway

The best method by far of getting around the city is by the Metro underground system. As of 2016/20, parts of the network are being upgraded, so closures are taking place at weekends and often during the night. Check for any warnings, and plan your days ahead.

The network does not just cover DC; several surrounding towns in Virginia and Maryland are also linked. Further bus routes are available at the extreme ends of each line. The lines are colour-coded and are vastly simpler than many subway systems.

Arlington Cemetery is usually a good place to begin a tour of DC. It is free, just a 100yds from the Arlington Metro station, and it leaves you in suspense for the exciting museum exhibits.

To use the Metro, a new system was introduced in 2015. At the ticket machines, purchase a travel card called 'SmartTrip.' As of 2020, it costs a mere $2 to purchase the card. You then use 'up' or 'down' buttons to add credit onto it. This is to cover your journeys that day and /or beyond, and does not expire. If your credit runs out before leaving a station, the exit turnstile will electronically give you a rude message to top it up at another machine before allowing you to exit. It is a very simple system, but was a surprise and added confusion when we were faced with it the first time.

More info is available on the Washington Metro website:

Online Trip Planner - Washington Metro
(https://wmata.com/schedules/trip-planner/index.cfm)

Washington Metropolitan Area Transit Authority SmarTrip Card
(https://wmata.com/fares/buy-SmarTrip-card.cfm)

Bicycle

Some US cities operate a bike rental system. For a relatively small membership fee, you can rent a bike from a rack and ride it to another. In Washington DC, a daily fee of $8 (as of 2020) allows any bike to be ridden to any other bike station, checked in at the rack whilst you look around the area before hiring another bike for the next leg of your journey. Providing each trip is no more than 30 minutes, no other charge is made.

Several hundred bike stations like this one exist in strategic parts of the city. Borrow a bike for less than 30 minutes and return it to any other rack. Continue all day for $8 per person.

Apply for membership online: https://www.capitalbikeshare.com/

An annual pass is available if you are perhaps planning several visits in one year. A brochure can be obtained at the racks with a map of all the bike stations (marked with red squares).

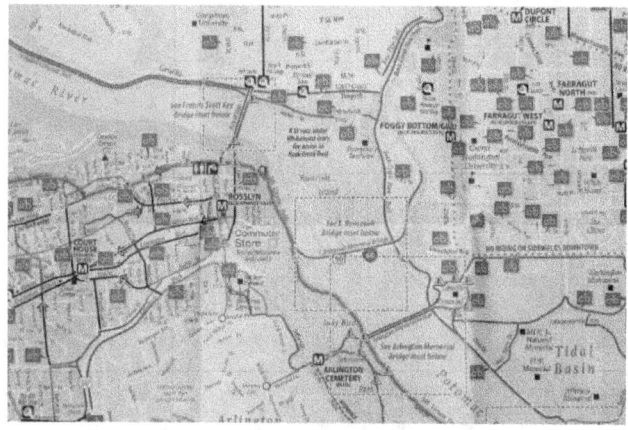

Tour Bus

A hop-on hop-off tour bus service is available throughout the main part of the city. Tickets are purchased in three available tiers, and although it is a little on the expensive side for my personal liking, it is still a great service.

The latest information and prices can be found here: www.trolleytours.com

There are twenty-five points at strategic locations across the city where you can access the service. One ticket allows unlimited rides to all points from 9am to 5pm. Special tours also include a vehicle that takes you along the Pontiac River – yes you read that correctly. From driving along the road to sailing down the river – without changing vehicle. Another specialised tour is a night excursion around the floodlit monuments.

Another bus company is www.bigbustours.com which is more expensive but can include free entrance to the Madame Tussauds museum: (www.madametussauds.com/washington-dc)
It provides forty potential stops instead of twenty-five, a two-day pass option, and also night tours.

Chapter 3   Arlington Cemetery

www.arligtoncemetery.mil

Found just to the West of Washington DC, many astronauts and other spaceflight-related personnel are buried at Arlington Cemetery. These are listed in alphabetical order along with a brief history of their spaceflight connection. The interest in seeking specific graves may sound to some a little morbid, but if you never had the chance to meet these astronauts – who represent such an incredible part of space history – while they were alive, then this is the only option left. It is also a celebration of the work and memories of these brave explorers of the cosmos, and I feel humbled just looking at their headstones. The artwork on some of the graves is in itself also of great interest.

To visit a specific plot, a map can be drawn out at the information desk by, for example, looking up an astronaut's name on a dedicated computer terminal, (found near the desk), or by downloading the App here:

ANC Explorer (Apple App Store)
Or here:
ANC Explorer (Google Play)

This will guide you around the cemetery straight to the plot. Details are available at the Visitor Center or on the website above.

There are several methods to obtain an accurate map to show where each grave is situated. Type in the deceased's full name on this machine, and follow the instructions. Other methods include a free phone 'App' or just ask at the desk in the Visitor Center.

This and all the grave images are by the author.

The Cemetery itself is not actually geographically located within Washington DC, but is today in the State of Virginia. When the original City was mapped out, the land the Cemetery now occupies was then a part of DC. Just before the Civil War, during an active slavery trade, which took place on the west of the Potomac River, the Union supporting politicians decided to remove the association by moving the DC border to the east of the River. The District of Columbia was no longer a diamond as originally intended.

Tips:
Arlington Cemetery is a necessary see place to visit, though never underestimate the size of the site and number of graves! Do not expect to turn up and wander around at random to see the graves or monuments on your list. It *does* require planning and research, or you will almost certainly be making a wasted journey. Do respect the site – it is not for picnics or a children's playground. Wardens are on duty to monitor any disrespectful activities. At the JFK grave, do remove hats and sunglasses, (you will be asked to do so anyway).

Take plenty of drinking water. No food is allowed in the Cemetery grounds, but if you require a nibble for medical reasons, be discreet. To conserve energy, plan each grave visit in a logical geographical sequence, and take a hat and sunglasses as there is little shelter from the Sun. Do not underestimate the amount of walking and time required for your visit. The map shows clearly numbered areas, but paths can cut across each region and become very confusing. To the far west of the site, the name of each road is marked on the map, but not always on the road itself. Use the area numbers and curves of each road as a clue as to where you are located compared to the desired gravesite you wish to visit.

Grave-numbers are marked on the back of each headstone. They *do* run consecutively but in a very odd way for each row. Systems have changed over the decades, and plot borders are sometimes shown by a change in the numbering sequence, so look for a row in front or behind to continue searching for a particular grave.

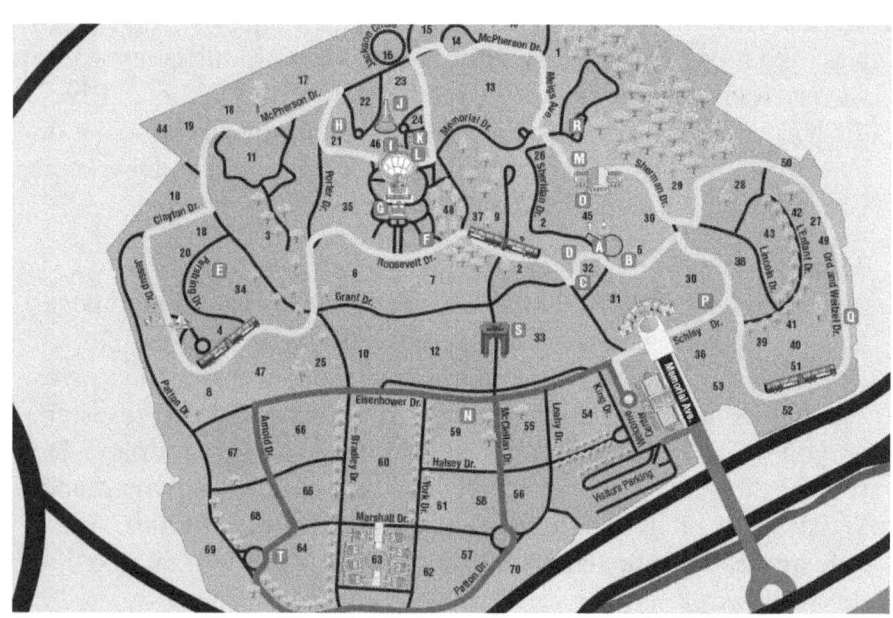

Never underestimate the sheer number of graves and the distance you may need to walk. Conserve your energy by carefully planning your day.

13

Michael P. Anderson, Lieutenant Colonel, US Air Force
(1959-2003) *Plot: Section 46, Lot 1180-1*
*Just behind the Shuttle Memorial

Anderson graduated from the University of Washington in 1981 and was commissioned a second lieutenant. After completing a year of technical training at Keesler Air Force Base, (AFB), Mississippi, he was assigned to Randolph AFB Texas.

In 1986, he was selected to attend Undergraduate Pilot Training at Vance AFB, Oklahoma. Upon graduation he was assigned to the 2nd Airborne Command and Control Squadron, Offutt AFB Nebraska as an EC 135 pilot, flying the Strategic Air Commands airborne command post, code-named 'Looking Glass'. From January 1991 to September 1992 he served as an aircraft commander and instructor pilot in the 920th Air Refuelling Squadron, Wurtsmith AFB Michigan. Anderson logged over 3000 hrs in the KC-135 and the T-38A aircraft.

STS-89 Endeavour (January 22-31, 1998) was the eighth Shuttle-Mir docking mission during which the crew transferred more than

9,000 pounds of scientific equipment, logistical hardware, and water from the Space Shuttle to Mir.

In the fifth and last exchange of a U.S. astronaut, STS-89 delivered Andy Thomas to Mir and returned with David Wolf. Mission duration was 8 days, 19 hours and 47 seconds, travelling 3.6 million miles in 138 orbits.

STS-107 Columbia (16 January to 1 February 2003). The 16-day flight was a dedicated science and research mission. Working 24 hours a day, in two alternating shifts, the crew successfully conducted approximately 80 experiments. The mission ended abruptly on 1st February 2003 when Space Shuttle Columbia and the crew perished during re-entry, 16 minutes before they were scheduled to land. Mission duration was 15 days, 22 hours and 20 minutes.

Michael's grave is located behind the Columbia Memorial.

Charles Arthur Bassett II - US Air Force (1931-1966)
Plot: Section 4, Lot 195 Grid AA-11.5
Charles Bassett was an Air Force Captain (no relation of mine as far as I am aware). He graduated from the Aerospace Research

Pilot School and the Air Force Experimental Pilot School. He then served as an experimental test pilot in the Fighter Projects Office at Edwards Air Force Base, California, and logged over 3,600 hrs-flying time, including over 2,900 hours in a jet aircraft. Charles was one of the third group of astronauts named by NASA in October 1963.

In addition to participating in the overall astronaut-training program, he had specific responsibilities pertaining to training and simulators. On 8 November 1965, he was selected as pilot of the upcoming Gemini 9 mission. He died on 28 February 1966, in the crash of a T-38 jet trying to land in dense fog at the Johnson Space Center, Houston, Texas, along with Elliott See, also buried alongside him.

Alan LaVern Bean, US Navy (1932-2018)
Plot: Section 11 grave 2492-B
American Astronaut. As a member of NASA's Apollo 12 program, he became the fourth man to walk on the surface of the Moon (1969). He attained his Bachelor of Science degree from the University of Texas and following graduation, he was trained

as a fighter pilot with the United States Navy. After several years as a test pilot, he joined NASA in 1963 and went on to serve as a backup crewman for the Gemini 10 (1966) and later Apollo 9 (1969) missions.

On November 19, 1969, Bean served as the Lunar Module pilot on the Apollo 12 mission commanded by Pete Conrad and Richard Gordon. Conrad and Bean respectively, became the third and fourth men to walk on the Moon. In July 1973, Bean served as commander of the second crew to inhabit the Skylab and over a 59 day period, important data and experiments were conducted. Bean retired from the Navy in 1975 and remained with NASA until his retirement in 1981. He went on to become an accomplished painter who created works based around his space expeditions.

David M. Brown, Captain, US Navy (1956-2003)
Plot: Section 46 Lot 1180-3 *Just behind the Shuttle Memorial
David Brown joined the Navy after his internship at the Medical University of South Carolina. Upon completion of flight surgeon training in 1984, he reported to the Navy Branch Hospital in Adak, Alaska, as Director of Medical Services. He was then assigned to Carrier Airwing Fifteen, which deployed aboard the USS Carl

Vinson in the western Pacific. In 1988, he was the only flight surgeon in a ten-year period to be chosen for pilot training. Brown was designated a naval aviator in 1990 in Beeville, Texas, ranking number one in his class, and attended training and carrier qualification in the A-6E Intruder. In 1991, he reported to the Naval Strike Warfare Center in Fallon, Nevada, where he served as a Strike Leader Attack Training Syllabus Instructor and a Contingency Cell Planning Officer. Additionally, he was qualified in the F-18 Hornet, and deployed from Japan in 1992 aboard the USS Independence. In 1995, he reported to the U.S. Naval Test Pilot School as their flight surgeon, where he also flew the T-38 Talon. Brown logged over 2,700 flight hours with 1,700 in high performance military aircraft, and was qualified as a pilot in a NASA T-38 aircraft.

Selected as an astronaut in April 1996, Brown reported to the Johnson Space Center in August 1996. He completed two years of training, and was qualified as a mission specialist. He was initially assigned to support payload development for the International Space Station, followed by the astronaut support team responsible for Orbiter cockpit set-up, crew strap-in, and landing.

Brown was qualified as a pilot in a NASA T-38 aircraft. Selected as an astronaut April 1996, he reported to the Johnson Space Center in August 1996. He completed two years of training, and was qualified as a mission specialist. He was initially assigned to support payload development for the International Space Station, followed by the astronaut support team responsible for Orbiter cockpit set-up, crew strap-in, and landing.

Dave Brown flew aboard STS-107, logging 15 days, 22 hours, and 20 minutes in space. The 16-day flight was a dedicated science and research mission. The crew successfully conducted approximately 80 experiments. The STS-107 mission ended abruptly on February 1, 2003 when Space Shuttle Columbia and the crew perished during entry, 16 minutes before scheduled landing.

Roger Bruce Chaffee - US Navy (1935-1967)
Plot: Section 3, Lot 2502-F, Grid Q-15/16
Roger Chaffee, a United States Navy Lieutenant Commander, entered the Navy in 1957. He served as safety officer and quality control officer for Heavy Photographic Squadron 62 at the Naval Air Station in Jacksonville, Florida.

In January 1963, he entered the Air Force Institute of Technology at Wright-Patterson Air Force Base, Ohio, to work on a Master of Science Degree in Reliability Engineering. He logged more than 2,300 hrs flying time, including more than 2,000 hours in jet aircraft.

Chaffee was one of the third group of astronauts selected by NASA in October 1963. In addition to participating in the overall training program, he was tasked with working on flight control communications systems, instrumentation systems, and attitude and translation control systems in the Apollo Branch of the Astronaut office.

On 21 March 1966 he was selected as one of the pilots for the AS-204 mission, the first 3-man Apollo flight. Lieutenant Commander Chaffee died on 27 January 1967 in the Apollo spacecraft flashfire during a launch pad test at Kennedy Space Center, Florida, also killing Gus Grissom, and Ed White, (who is buried at West Point, New York).

Laurel Blair Salton Clark, Captain, US Navy (1964-2003)

*Plot: Section 46, Lot 1180-2 *Just behind the Shuttle Memorial*
Laurel Blair Clark graduated from William Horlick High School, Racine Wisconsin in 1979. She received Bachelor of Science degree in zoology from the University of Wisconsin-Madison in 1983, and a doctorate in medicine from the same school in 1987. She was a member of the Aerospace Medical Association, and Society of US Naval Flight Surgeons.

Selected by NASA in April 1996, Dr Clark reported to the Johnson Space Center in August 1996. After completing two years of training and evaluation, she was qualified for flight assignment as a mission specialist. From July 1997 to August 2000, Dr Clark worked in the Astronaut Office Payloads / Habitability Branch.

During medical school, she did active duty training with the Diving Medicine Department at the Naval Experimental Diving Unit in

March 1987. After completing medical school, Dr Clark underwent postgraduate Medical education in Paediatrics from 1987-1988 at Naval Hospital Bethesda, Maryland. The following year she completed Navy undersea medical officer training at the Naval Undersea Medical Institute in Groton Connecticut, and diving medical officer training at the Naval Diving and Salvage Training Center in Panama City, Florida.

She was assigned as the Submarine Squadron Fourteen Medical Department Head in Holy Loch, Scotland. During that assignment she dove with US Navy divers and Naval Special Warfare Unit Two Seals, and performed numerous medical evacuations from US submarines. After two years of operational experience, she was designated as a Naval Submarine Medical Officer and Diving Medical Officer. She underwent 6 months of aero-medical training at the Naval Aerospace Medical Institute in Pensacola, Florida, and was designated as a Naval Flight Surgeon. She was stationed at MCAS Yuma, Arizona and assigned as Flight Surgeon for a Marine Corps - Night Attack Harrier Squadron (VMA 211). She made numerous deployments, including one overseas to the Western Pacific. Her squadron won the Marine Attack Squadron of The Year for its successful deployment. She became the Group

Flight Surgeon for the Marine Aircraft Group (MAG 13). Prior to her selection as an astronaut candidate, she served as a Flight Surgeon for the Naval Flight Officer advanced training squadron (VT-86) in Pensacola, Florida.

Dr Clark flew aboard STS-107, logging 15 days, 22 hours, and 20 minutes in space. The 16-day flight was a dedicated science and research mission. The crew successfully conducted approximately 80 experiments. The mission ended abruptly on February 1, 2003 when Space Shuttle Columbia and the crew perished during re-entry, 16 minutes before scheduled landing.

Charles 'Pete' Conrad, Captain, US Navy (1930-1999)
Plot: Section 11, Site 113-3
Following graduation from Princeton University in 1953, Pete Conrad entered the Navy and became a naval aviator. He then attended the Navy Test Pilot School at Patuxent River, Maryland, where he became a Project Test Pilot. Conrad also served as a flight instructor and performance engineer at the Test Pilot School, and after completing his tour of duty at Patuxent River, he served as instructor pilot in F4H Phantoms on VF-121 and was assigned duty in VF-96 on board USS Ranger.

In September of 1962, NASA selected Conrad as an astronaut. His first flight was Gemini V, which established the space endurance record and placed the United States in the lead for person-hours in space. As commander of Gemini XI, Conrad helped to set a world altitude record. He then served as commander of Apollo XII, the second lunar landing, and was the third man to walk on the Moon. On his final space mission, he served as commander of Skylab II, the first US Space Station.

In December 1973, after serving 20 years, (11 of which were as an astronaut in the space program), Conrad retired from the U.S. Navy to accept a position as Vice President, Operations and Chief Operating Office of American Television and Communications Corporation (ATC). He was responsible for both the operation of existing systems and the national development of new cable television systems. In 1976, he resigned from ATC to accept the position of Vice President and consultant to McDonnell Douglas Corporation. In 1978, he became Vice President of marketing and was responsible for all commercial and military sales. Conrad then became Senior Vice President, Marketing in 1980. He was appointed as Product Support President in 1982 and 1984, and was named Staff Vice President of International Business Development.

In 1990, Conrad became Staff Vice President, New Business for McDonnell Douglas Space Company, where he participated in research and development for the Space Exploration Initiative. Included for research and development are the construction of the then Space Station Freedom, the return to and colonisation of the Moon, and the exploration of Mars. He contributed his expertise on the Single-Stage-To-Orbit and return space transportation system called the Delta Clipper. In 1993 Conrad became Vice President of Project Development. Conrad died 8 July 1999 from injuries sustained in a motorcycle accident in Ojai, California.

Donn F. Eisele - US Air Force (1930-1987)
Plot: Section 3, Lot 2503-G-1, Grid Q-15
Donn Eisele graduated from the United States Naval Academy and chose a career in the Air Force. He is also a graduate of the Air

Force Aerospace Research Pilot School at Edwards Air Force Base, California.

He was a project engineer and experimental test pilot at the Air Force Special Weapons Center at Kirtland Air Force Base, New Mexico. In this capacity, he flew experimental test flights in support of special weapons development programs. He logged more than 4,200 hours flying time – 3,600 hours in jet aircraft.

Eisele was one of the third group of astronauts selected by NASA in October 1963. On 11 October 1968, he occupied the command module pilot seat for the eleven-day flight of Apollo 7 – the first manned flight test of an Apollo. With spacecraft commander Walter M. Schirra, Jr., and lunar module pilot Walter Cunningham, Eisele participated in and executed manoeuvres. This task enabled the crew to perform exercises in transposition and docking and lunar orbit rendezvous with the S-IVB stage of their Saturn IB vehicle. They also completed eight successful test and manoeuvring ignitions of the service module propulsion engine, measured the accuracy of performance of all spacecraft systems, and provided the first effective television transmissions of onboard crew activities.

Apollo 7 entered Earth-orbit with an apogee, (furthest point from Earth), of 153.5 nautical miles, and perigee, (closest point to Earth), of 122.6 nautical miles; and the 260hr / 4, 500,000 miles shakedown flight was successfully ended on 22 October 1968. Splashdown occurred in the Atlantic, eight miles from the aircraft carrier, Essex, (only ⅓ of a mile from the originally predicted aiming point). Eisele served as backup command module pilot for the Apollo 10 flight.

In July 1972, Colonel Eisele retired from the Air Force and left the space program to become Director of the U.S. Peace Corps in Thailand. Upon Returning from Thailand, Eisele became Sales Manager for Marion Power Shovel Company. He handled private and corporate accounts for the investment firm Oppenheimer & Company.

Theodore Cordy Freeman - US Air Force (1930-1964)
Plot: Section 4 Lot 3148 Grid AA-11

Theodore Freeman completed his secondary education in 1948. He attended the University of Delaware at Newark for one year, then entered the United States Naval Academy, and graduated in 1953 with a Bachelor of Science degree. In 1960, he received a Master of Science degree in Aeronautical Engineering from the University of Michigan, and was a member of the American Institute of Aeronautics and Astronautics, and the Society of Experimental Test Pilots.

Freeman graduated from both the Air Force's Experimental Test Pilot and Aerospace Research Pilot Courses. He elected to serve with the Air Force. His last Air Force assignment was as a flight test aeronautical engineer and experimental flight test instructor at the Aerospace Research Pilot School at Edwards Air Force Base, California. He served primarily in performance flight-testing and stability testing areas. He logged more than 3,300 hours flying time, including more than 2,400 hours in jet aircraft.

Freeman died on 31 October 1964, at Ellington Air Force Base, Houston, Texas, in a T-38 jet crash. Some of the other astronauts in the group at the time agreed that he had all the qualities to become the first man on the Moon.

THEODORE C
FREEMAN
DELAWARE
CAPTAIN
US AIR FORCE
FEB 18 1930
OCT 31 1964

John Glenn – (1921 – 2016)
Section 35 Grave 1543

John Herschel Glenn was the first American to orbit the Earth. He was a veteran of both World War II and the Korean War, serving as a fighter pilot in the Navy and Marine Corps. After the Korean War ended, he became a test pilot, and when the newly formed NASA began recruiting astronauts in 1958, he applied and was selected as one of the original Mercury 7.

On 20 February 1962, he became the third American in space and the first to orbit the Earth when he lifted off in Friendship 7, which is on display at the Air & Space Museum, Washington DC. He left NASA in 1964 and retired from the Marines a year later. He first entered the world of politics by running for the US Senate from Ohio in 1964, but had to withdraw from the race early due to a concussion sustained during a fall.

He ran again ten years later in 1974 for the Senate and became a Senator. He served for four terms until retiring in 1999. He was the chief author of the Nuclear Non-Proliferation Act of 1978, and sought the Democratic presidential nomination in 1984. Before retiring fully, at the age of 77 he flew on Space Shuttle Discovery in 1998, becoming the oldest human to fly in space. In his later years, he founded the John Glenn Institute of Public Service at Columbus, Ohio. He also taught at the school as an adjunct professor.

Stanley David Griggs – Rear Admiral, US Navy (1939-1989)
Plot: Section 7A, Lot 81 Grid T/U-23
Stanley Griggs graduated from Annapolis in 1962 and entered pilot training shortly thereafter. In 1964, he received his Navy wings and was attached to Attack Squadron-72 flying A-4 aircraft. He completed one Mediterranean cruise and two Southeast Asia combat cruises aboard the aircraft carriers USS Independence and USS Roosevelt. Mr. Griggs entered the U.S. Naval Test Pilot School at Patuxent River, Maryland, in 1967, and upon completion of test pilot training, was assigned to the Flying Qualities and Performance Branch, Flight Test Division, where he flew various test projects on fighter and attack-type aircraft. In 1970, he resigned his regular United States Navy commission and affiliated with the Naval Air Reserve in which he held the rank of Rear Admiral.

He logged 9,500 hours flying time – 7,800 hours in jet aircraft – and flew more than 45 different types of aircraft, including single and multi-engine prop, turbo prop and jet aircraft, helicopters, gliders, hot air balloons, and the Space Shuttle. He made over 300 carrier landings, held an airline transport pilot license, and was a certified flight instructor.

In July 1970, Griggs was employed at the Lyndon B. Johnson Space Center, Texas, as a research pilot, working on various flight test and research projects in support of NASA programs. In 1974, he was assigned duties as the project pilot for the shuttle trainer aircraft, and participated in the design, development and testing of those aircraft, pending their operational deployment in 1976. He was appointed Chief of the Shuttle Training Aircraft Operations Office in January 1976. He was responsible for the operational use of the shuttle trainer, and held that position until selected as an astronaut candidate by NASA in January 1978. In August 1979, he completed a training evaluation period and became eligible for Space Shuttle flight crew assignment.

From 1979 to 1983, Griggs was involved in several Space Shuttle engineering capacities, including the development and testing of the Head-Up Display (HUD) approach and landing avionics system, the Manned Manoeuvring Unit, (MMU), and the requirements definition and verification of in-orbit rendezvous and entry flight phase software and procedures. In September 1983, he began crew training as a mission specialist for flight STS 51-D, which flew 12-19 April 1985. During the flight, Griggs conducted the first unscheduled extravehicular activity, (EVA, or 'spacewalk'), of the space program. It lasted for over three hours, during which preparations for a satellite rescue attempt were completed.

At the time of his death, Mr. Griggs was in flight crew training as pilot for STS-33, a dedicated Department of Defence mission, scheduled for launch in August 1989. He died on 17 June 1989, near Earle, Arkansas, in the crash of a vintage World War II airplane.

Virgil Ivan Grissom - US Air Force (1926-1967)
Plot: Section 3, Lot 2503-E, Grid Q-15/16

Virgil 'Gus' Grissom, Air Force Lieutenant Colonel, received his wings in March 1951. He flew 100 combat missions in Korea in F-86s with the 334th Fighter Interceptor Squadron. Upon returning to the US in 1952 he became a jet instructor at Bryan, Texas.

In August 1955, he entered the Air Force Institute of Technology at Wright-Patterson Air Force Base, Ohio, to study Aeronautical Engineering. He attended the Test Pilot School at Edwards Air Force Base, California, in October 1956, and returned to Wright-Patterson in May 1957 as a test pilot. He logged 4,600 hours flying time, including 3,500 hours in jet aircraft.

Grissom was one of the seven Mercury astronauts selected by NASA in April 1959. He piloted the Liberty Bell 7 spacecraft – the second and final sub-orbital Mercury test flight – on 21 July 1961. This flight lasted 15 minutes and 37 seconds, attained an altitude of 118 statute miles, and travelled 302 miles downrange from the launch pad at Cape Kennedy.

On 23 March 1965, he served as command pilot on the first manned Gemini flight, a 3-orbit mission during which the crew accomplished the first orbital trajectory modifications and the first lifting re-entry of a manned spacecraft. He also became the very first astronaut to enter space twice. Grissom was named to serve as command pilot for the AS-204 mission, the first 3-man Apollo flight, better known as 'Apollo 1'. Lieutenant Colonel Grissom died on 27 January 1967 in the Apollo 1 spacecraft flashfire during a launch pad test at Kennedy Space Center, Florida.

Conspiracy link: some say that Grissom and the other two members of the Apollo 1 crew were murdered because they were supposedly about to reveal to the public that the Moon landings were going to be hoaxed. We have a book out on that too.

James Benson Irwin - US Air Force (1930-1991)
Plot: Section 3 Lot 2503-G-2 Grid Q-15.5
James Irwin, an Air Force Colonel, was commissioned in the Air Force upon graduation from the Naval Academy in 1951. He received flight training at Hondo and Reese Air Force Base, Texas.

Prior to reporting for duty at the Manned Spacecraft Center (MCC) Houston, he was assigned as Chief of the Advanced Requirements Branch at the Air Defense Command. He graduated from the Experimental Test Pilot School in 1961 and from the Air Force Aerospace Research Pilot School in 1963.

He also served with the F-12 Test Force at Edwards Air Force Base, California, and with the AIM 47 Project Office at Wright-Patterson Air Force Base, Ohio. During his military career, he accumulated more than 7,015 hours in flight, including 5,300 hours in jet aircraft.

In 1961, his aircraft crashed on a training flight, leaving him with two broken legs, a broken jaw, and a concussion that wiped out his memory. It took extensive psychiatric treatment and hypnosis to restore the lost memory and enable him to resume flying 14 months later. A potential heart problem was noted but he managed to keep it quiet to resume his flying career.

Colonel Irwin was one of the 19 astronauts selected by NASA in April 1966. He was crew commander of lunar module (LTA-8) – this vehicle finished the first series of thermal vacuum tests on 1st June 1968. He also served as a member of the astronaut support crew for Apollo 10, and as backup lunar module pilot for the Apollo 12 flight.

Irwin served as lunar module pilot for Apollo 15, 26 July to 7 August 1971. His companions on the flight were David R. Scott, spacecraft commander, and Alfred M. Worden, command module pilot. It was the fourth manned lunar landing mission. They explored the moon's Hadley Rille and Apennine Mountains, which are located on the southeast edge of the Mare Imbrium (Sea of Rains). The lunar module 'Falcon' remained on the lunar surface for 66 hours, 54 minutes, setting a new record for lunar surface stay time, and Scott and Irwin logged 18 hours and 35 minutes each in extravehicular activities conducted during three separate excursions across the lunar surface.

Using the Lunar Roving Vehicle, (LRV or 'Rover'), to transport themselves and their equipment along portions of Hadley Rille and the Apennine Mountains, Scott and Irwin performed a selenological inspection and survey of the area, and collected approximately 180 pounds of lunar surface materials. They deployed an ALSEP package, which involved the emplacement and activation of surface experiments, and their lunar surface activities were televised in colour. Ground controllers stationed in Houston, Texas, operated this camera remotely. Other mission achievements included: largest payloads ever placed in Earth and lunar orbits; first scientific instrument module bay flown and operated on an Apollo spacecraft; longest distance traversed on lunar surface; first use of a lunar surface navigation device, mounted on the Rover, and the first sub-satellite launched in lunar orbit. Finally, the mission included the first EVA, (spacewalk), from a command module during trans-Earth coast. Al Worden accomplished the latter feat during three spacewalks to command module Endeavour's SIM bay, where he retrieved film cassettes from the panoramic and mapping cameras. Apollo 15 concluded with a Pacific splashdown and recovery by the USS Okinawa.

In completing his first flight, Irwin logged 295 hours and 11 minutes in space – 19 hours and 46 minutes of which were in EVA. Colonel Irwin resigned from NASA and the Air Force in July 1972 to form a religious organisation, High Flight Foundation, in Colorado Springs, Colorado.

On two occasions, he led expeditions to Mount Ararat in Turkey in search of evidence of Noah's Ark. In 1982, he reached the 16,946-foot summit but fell on the glacier, suffering severe leg and face lacerations. He had to be carried down on horseback.
A year later, he surveyed the summit by airplane, looking for possible remains of the ark, which, according to the Book of Genesis, came to rest on the 'Mountains of Ararat'.
"It's easier to walk on the moon," Irwin said. "I've done all I possibly can, but the ark continues to elude us."

In 1990, Irwin travelled to the UK and gave a series of lectures. During the tour, he opened a public observatory in Canterbury, Kent, which is named after him. The author of this book helped to acquire it for the Mid-Kent Astronomical Society. James Irwin died of a heart attack on 8 August 1991.

Image www.midkentastro.org.uk

Astronaut Clifton Williams is buried just one row behind him.

Brian Joseph James – Major, US Marine Corps (1957-1992)

(Plot number 60 4436 – please get in touch if a reader finds Grid Number)

A Baltimore, Maryland native, Brian James graduated from the Baltimore Polytechnic Institute in 1975. He received an Associate of Arts degree from the Community College in Baltimore in 1977. In 1979, he received a Bachelor's degree in Political Science from the University of Maryland. He was commissioned a Second

Lieutenant in the Marine Corps in 1980, and married Deanna Lynn Batton in 1982.

In 1988, he was reassigned to the Naval Air Station at Whiting Field in Florida as a flight instructor. He earned a Master's Degree in Mathematics from the University of West Florida in 1989. He later worked at the Naval Air Station at Patuxent River, where he graduated from the Naval Test Pilot School, Class 99, in June 1991 as an engineering test pilot. In 1984, he served in operations in Beirut, Lebanon. He was serving with the Rotary Wing Aircraft Test Directorate when he died making a landing at the Marine Corps Air Station at Quantico, Virginia. The Marine Corps had also selected him as an Astronaut Candidate.

Brian Joseph James, 34, a Marine Corps Major who died 20 July 1992 in the crash of the V-22 Osprey aircraft into the Potomac River, Fort Myer, Virginia. Burial was with full military honours.

Iven Carl Kincheloe, Jr. - US Air Force (1928-1958)
Plot: Section 2, Lot 4872-1 Grid V-32.5 (Not far from JFK)
Iven Kincheloe was a pioneer astronaut. Born in Detroit, Michigan, he was a natural pilot, flying solo on his 16th birthday, and in fact went on to be America's first 'spaceman'. He became a 2nd

Lieutenant in the US Air Force in 1949. During the Korean War he flew a F-86E Sabre Jet and was a double ace. After the war, he was promoted to Captain and became a test pilot at Edwards Air Force Base, California. In the early 1950s at Edwards he test-flew the new fighters McDonnell F-101, Convair F-102, Lockheed F-104 and Republic F-105.

Kincheloe joined the research team on the Bell X-2 aircraft project in May 1956, with the goal to reach an altitude of 100,000ft. On 7 September 1956, Kincheloe piloting the Bell X-2 reached the top of its curve at 126,500ft above the Earth's atmosphere, making him the first man to reach space – 5 years before Yuri Gagarin, (who was first to orbit the Earth).

The legal height limit for an aircraft at the time was 100,000ft, beyond which aircraft are not designed to function in their usual way due to the tenuous atmosphere, but instead must operate as spacecraft. Kincheloe was subsequently selected to be pilot of the more powerful X15 rocket-plane, which was capable of altitudes of up to 102 000 feet, but tragically lost his life when his F-104 crashed on take-off at Edwards Air Base on 26 July 1958.

Robert Francis Overmyer, Colonel, US Marines (1935-1996)
Plot: Section 23 Grave 22469

Robert Overmyer entered active duty with the Marine Corps in January 1958. After completing Navy flight training in Kingsville, Texas, he was assigned to Marine Attack Squadron 214 in November 1959. He then attended the Naval Postgraduate School in 1962 to study aeronautical engineering. Upon completion of his graduate studies, he served one year with Marine Maintenance Squadron 17 in Iwakuni, Japan, and was then transferred to the Air Force Test Pilots School at Edwards Air Force Base, California. He was chosen as an astronaut for the USAF Manned Orbiting Laboratory Program in 1966. However, the program was cancelled in 1969. During this time he gained over 7,500 flight hours with over 6,000 in jet aircraft.

Overmyer was selected as a NASA astronaut in 1969 after the MOL Program was cancelled. His first assignment with NASA was engineering development duties on the Skylab Program from 1969 until November 1971. From November 1971 until December 1972, he was a support crewmember for Apollo 17, and was the launch capsule communicator. From January 1973 until July 1975 he was a support crewmember for the Apollo-Soyuz Test Project, and was the NASA capsule communicator in the Mission Control Center in Moscow, USSR. In 1976, he was assigned duties on the Space Shuttle Approach and Landing Test (ALT) Program, and was the prime T-38 chase pilot for Orbiter Free-Flights 1 and 3. In 1979, Overmyer was assigned as the Deputy Vehicle Manager of the Columbia, in charge of finishing the manufacturing and tiling of the Orbiter at the Kennedy Space Center for its first flight in April 1981.

Colonel Overmyer was the pilot for STS-5, the first fully operational flight of the Shuttle, which launched from Kennedy Space Center, Florida on 11 November 1982. Spacecraft commander, Vance D. Brand, and two mission specialists, Dr Joseph P. Allen and Dr William B. Lenoir, accompanied him. This was the very first mission in history with a four-man crew, clearly demonstrating the Shuttle as operational by the successful first deployment of two commercial communications satellites from the payload bay. The mission marked the first use of the Payload Assist Module, (PAM-D), and its new ejection system. Numerous flight tests were performed throughout the mission to document Shuttle performance during launch, boost, orbit, atmospheric re-entry, and landing phases. It was the last flight to carry the Development Flight Instrumentation (DFI) package to support flight-testing. A Getaway Special, three Student Involvement Projects and medical experiments were included on the mission. The crew concluded the 5-day orbital flight of Columbia with the first entry and landing through a cloud deck to a hard-surface runway, and demonstrated maximum braking. Mission duration was 122 hours before landing on the runway at Edwards Air Force Base, California, on 16 November 1982.

Overmyer was the commander of STS 51-B, the Spacelab-3 (SL-3) mission. He commanded a crew of four astronauts and two payload specialists conducting a broad range of scientific experiments, ranging from space physics to the suitability of animal holding facilities. Mission 51-B was also the first Shuttle flight to launch a small payload from the 'Getaway Special' canisters. It was launched at 12:02pm EDT on 29 April 1985 from Kennedy Space Center, Florida and landed at Edwards Air Force Base, California on 6th May 1985. The mission completed 110 orbits of the Earth.

Colonel Overmyer retired from NASA and the Marine Corps in May 1986. He was killed in a crash in a light aircraft he was testing on 22 March 1996.

Stuart Allen Roosa, United States Air Force (1933-1994)
Plot: Section 7A, Lot 73, Grid T/U-23.5
Stuart Roosa was an experimental test pilot at Edwards Air Force Base, California, from September 1965 to May 1966, following graduation from the Aerospace Research Pilots School.

Stuart attended Gunnery School at Del Rio and Luke Air Force Bases. He was a graduate of the Aviation Cadet Program at Williams Air Force Base, Arizona, where he received his flight training commission in the Air Force. He logged 5,500 hours of flying time – 5,000 hours in jet aircraft.

Roosa was one of the 19 astronauts selected by NASA in April 1966. He was a member of the support crew for Apollo 9. He was Command Module pilot on Apollo 14, 31 January - 9 February

1971. With him on the third lunar landing mission were Alan B. Shepard (commander) and Edgar D. Mitchell (lunar module pilot). Manoeuvring their lunar module 'Antares' to a landing in the hilly upland Fra Mauro region of the moon, Shepard and Mitchell subsequently deployed and activated various scientific equipment and experiments, and collected almost 100 pounds of lunar samples. Throughout this 33-hour period of lunar surface activities, Roosa remained in lunar orbit aboard the command module 'Kittyhawk' to conduct a variety of assigned photographic and visual observations.

Apollo 14 achievements include first use of the Mobile Equipment Transporter (trolley); longest distance covered on the lunar surface; largest payload returned from the lunar surface; longest lunar surface time (33 hours), and longest lunar EVA (9 hours and 17 minutes). The mission also developed the use of shortened lunar orbit rendezvous techniques, first colour TV on the moon, as well as the first extensive orbital science period conducted during CSM (Command & Service Module) solo operations. Roosa logged a total of 216 hours and 42 minutes in space.

He served as backup command pilot for the Apollo 16 and 17 missions, and later worked on the development of the Shuttle program until his retirement in 1976.

From 5 February 1976 to 1 July 1977, Roosa served as Corporate Vice President of International Operations, U.S. Industries Inc. Illinois, and President, USI Middle East Development Company, Ltd., Athens, Greece. He initiated product development activities of appropriate divisions of U.S. Industries to insure product compatibility with Middle East countries. From July 1977 to March 1981 he was Vice President Advanced Planning, Charles Kenneth Campbell Investments and Commercial real estate development. From March 1981 he was president and owner of Gulf Coast Coors, Inc, Gulfport, Mississippi. Roosa died on 12 December 1994, due to complications of pancreatitis.

Francis Richard 'Dick' Scobee – Lieutenant Colonel, US Air Force (1939-1986) *Just beside the Shuttle Memorial.* Richard Scobee enlisted in the United States Air Force in 1957, trained as an engine mechanic, and was subsequently stationed at Kelly Air Force Base, Texas. While there, he attended night school and two years at college, which led to his selection for the Airman's Education and Commissioning Program. He graduated from the University of Arizona with a Bachelor of Science degree in Aerospace Engineering. He received his commission in 1965, and after receiving his wings in 1966 completed a number of assignments, including a combat tour in Vietnam. He returned to the United States and attended the Air Force Aerospace Research Pilot School at Edwards Air Force Base, California. After graduating in 1972, he participated in test programs, for which he flew such varied aircraft as the Boeing 747, X-24B, (transonic aircraft technology), F-111, and the C-5. He logged more than 6,500 hours flying time in 45 types of aircraft.

NASA selected Lt. Col. Scobee as an astronaut candidate in January 1978. In August 1979, he completed a training and evaluation period, making him eligible for assignment as a pilot on future space shuttle crews. In addition to astronaut duties, Scobee was an Instructor Pilot on the NASA/Boeing 747 shuttle carrier airplane. He first flew into space as pilot of STS 41-C, which launched from Kennedy Space Center on 6 April 1984. Crewmembers included spacecraft commander Captain Robert L. Crippen, and three mission specialists, Terry J. Hart, Dr G.D. (Pinky) Nelson, and Dr J.D.A. van Hoften. During this mission, the crew successfully deployed the Long Duration Exposure Facility. He retrieved the ailing Solar Maximum Satellite, repaired it onboard the orbiting Challenger, and replaced it in orbit using the robot arm. The mission also included flight-testing of Manned Manoeuvring Units in two spacewalks, operation of the Cinema 360 and IMAX Camera Systems, as well as a Bee Hive Honeycomb Structures student experiment. The mission lasted seven days before landing at Edwards Air Force Base, California, on 13 April 1984. With the completion of this flight, he logged a total of 168 hours in space.

Lt. Col. Scobee was commander on STS 51-L, which launched from Kennedy Space Center at 11:38:00 EST on 28 January 1986. The crew onboard the Orbiter Challenger included the pilot, M.J. Smith (U.S. Navy pilot), three mission specialists – Dr R.E.

McNair, Lt. Col. E.S. Onizuka (U.S. Air Force), and Dr J.A. Resnik, as well as two civilian payload specialists, G.B. Jarvis and S.C. McAuliffe. The STS 51-L crew died on January 28, 1986 when Challenger exploded shortly after launch. Scobee's grave is next to the Shuttle Challenger & Columbia memorials.

Regarding the graves of the Challenger disaster, one of the crew members' bodies was never recovered, and so personal items were placed in the grave instead. However, out of respect, it was never revealed which grave this applied to.

Elliott M. See, Jr - United States Navy (1927-1966)
Plot: Section 4, Lot 208, Grid T-9
Elliott See was a Naval Aviator from 1953 to 1956. He worked for the General Electric Company from 1949 to 1953 and 1956 to 1962 as a flight test engineer, group leader and experimental test pilot. He served as project pilot on the J79-8 engine development program in connection with F4H aircraft, and conducted flight tests on the J-47, J-73, J-79, CJ805 and CJ805 rear fan engines. This work involved flying in F-86, XF4D, F-104, F11F-1F, RB-66, F4H, and T-38 aircraft. See logged over 3,700 hours flying time, including 3,200 hours in jet aircraft.

See was one of nine pilot astronauts selected in September 1962. He participated in all phases of the astronaut-training program and was given responsibility for monitoring the design and development of guidance and navigation systems. He also aided in the coordination of mission planning. He was selected as pilot of the back-up crew for the Gemini 5 mission, and the command pilot for the Gemini 9 flight.

Elliott See died on 28 February 1966, in St. Louis, Missouri, in the crash of a T-38 aircraft along with Charles Bassett. They are buried next to each other.

Michael John Smith – Captain, US Navy (1945-1986)
Plot: Section 7A, Lot 208-1, Grid T/U-23.5

Michael Smith graduated from the United States Naval Academy in 1967 and subsequently attended the U.S. Naval Postgraduate School at Monterey, California. He completed Navy aviation jet training at Kingsville, Texas, receiving his aviator wings in May 1969. He attended the Advanced Jet Training Command (VT-21) where he served as an instructor from May 1969 to March 1971. During the two-year period that followed, he flew A-6 Intruders and completed a Vietnam cruise while assigned to Attack Squadron 52 aboard the USS KITTY HAWK (CV-63). In 1974, he completed U.S. Navy Test Pilot School and attended the Strike Aircraft Test Directorate at Patuxent River, Maryland, to work on the A-6E TRAM and CRUISE missile guidance systems. He returned to the U.S. Navy Test Pilot School in 1976 and completed an 18-month tour as an instructor. From Patuxent River, he was assigned to Attack Squadron 75 where he served as maintenance and operations officer while completing two Mediterranean deployments aboard the USS SARATOGA. He flew 28 different types of civilian and military aircraft, logging 4,867.7 hours of flying time.

Selected as an astronaut candidate by NASA in May 1980, he completed a one-year training and evaluation period in August 1981, qualifying him for assignment as a pilot on future Space Shuttle flight crews. He served as a commander in the Shuttle Avionics Integration Laboratory, Deputy Chief of Aircraft Operations Division, and Technical Assistant to the Director, Flight Operations Directorate, and was assigned to the Astronaut Office Development and Test Group.

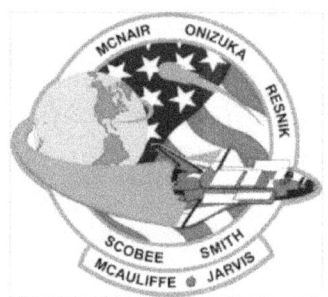

Captain Smith was assigned as pilot on STS 51-L, and was also selected as pilot for Space Shuttle Mission 61-N, scheduled for launch in the autumn of 1986. Captain Smith died on 28 January 1986, when the Space Shuttle Challenger exploded after launch from the Kennedy Space Center, also taking the lives of spacecraft commander Richard Scobee, three mission specialists, Dr R.E. McNair, Lieutenant Colonel E.S. Onizuka (USAF), and Dr J.A. Resnik, and two civilian payload specialists, Mr. G.B. Jarvis and Mrs. S. C. McAuliffe.

Stephen D. Thorne - US Navy (1953-1986)
Plot: Section 7A, Lot 135, Grid U-24

Upon graduation from the Naval Academy, Thorne entered flight training and received his wings in December 1976. Following training in the F-4 Phantom, he joined Fighter Squadron 21 (VF-21) and deployed to the Western Pacific aboard the USS Ranger. After training at the U.S. Naval Test Pilot School in 1981, Thorne spent the next two years at Strike Aircraft Test at the Naval Air Test Center, Patuxent River, Maryland, flying mostly ordnance and weapons systems tests in the F-4 and A-7 Corsair II. He completed F-18 Hornet transition training in October 1984 and joined Strike Fighter Squadron 132 (VFA-132) aboard USS Coral Sea until departing for NASA.

He accumulated over 2,500 hours and 200 carrier landings in approximately 30 different types of aircraft.

Thorne was selected as an astronaut by NASA in June 1985, and in August commenced a one-year training and evaluation program to qualify him for subsequent assignment as a pilot on future Space Shuttle flight crews. Lieutenant Commander Thorne was killed in an aircraft accident, in which he was a passenger, on 24 May 1986.

45

David Mathieson Walker, US Navy (1944-2001)
Plot: Section 66, Lot 5191

David Walker graduated from Annapolis and received flight training from the Naval Aviation Training Command at bases in Florida, Mississippi, and Texas. He was designated a naval aviator in December 1967, and proceeded to Naval Air Station Miramar, California, for assignment to F-4 Phantoms aboard the carriers USS Enterprise and USS America. From December 1970 to 1971, he attended the USAF Aerospace Research Pilot School at Edwards Air Force Base and became test pilot in the Naval Air Test Center. He then attended the U.S. Navy Safety Officer School at Monterey, California, and completed replacement pilot training in the F-14 Tomcat. In 1975, Walker was assigned to Fighter Squadron 142, stationed at Naval Air Station Oceana, Virginia as a fighter pilot, and was deployed to the Mediterranean Sea twice aboard the USS America. He has logged more than 7500 hours flying time – over 6500 hours in jet aircraft.

Selected by NASA in January 1978, Walker became an astronaut in August 1979. From July 1993 to June 1994, Walker was Chief of the Station/Exploration Support Office, Flight Crew Operations

Directorate, and then chaired the Johnson Space Center Safety Review Board.

A veteran of four space flights:
STS 51-A Discovery (8-16 November 1984) was launched from and returned to land at Kennedy Space Center, Florida. During the mission, the crew deployed two satellites, Canadas Anik D-2 (Telesat H), and Hughes LEASAT-1 (Syncom IV-1). In the first space salvage mission in history, the crew also retrieved for return to Earth the Palapa B-2 and Westar VI satellites. Mission duration was 127 Earth orbits in 7 days, 23 hours, 44 minutes.

STS-30 Atlantis (4-8 May 1989) was launched from Kennedy Space Center, Florida. During the 4-day mission, the crew successfully deployed the Magellan - Venus spacecraft, the first U.S. planetary science mission launched since 1978, and the first planetary probe to be deployed from the Shuttle. Magellan arrived at Venus in August 1990, and mapped over 95% of the surface. In addition, the crew also worked on secondary payloads involving fluid research in general, chemistry, and electrical storm studies observing 'Sprites'. Following 64 orbits of the Earth, the mission concluded with the first crosswind landing test of the Shuttle at Edwards Air Force Base, California.

STS-53 Discovery (2-9 December 1992) was launched from the Kennedy Space Center, Florida, and landed at Edwards Air Force Base, California. During 115 Earth orbits, the five-man crew deployed a classified Department of Defense payload DOD-1, and then performed several Military-Man-in-Space and NASA experiments. Mission duration was 175 hours, 19 minutes.

STS-69 Endeavour (7-18 September 1995) was launched from and returned to land at Kennedy Space Center, Florida. During the mission, the crew successfully deployed and retrieved a SPARTAN satellite and the Wake Shield Facility. Also on board was the International Extreme Ultraviolet Hitchhiker payload, numerous secondary payloads, and medical experiments. Mission duration was 10-days, 20 hours, 28 minutes. David died of cancer on 23 April 2001.

Clifton C. Williams, Jr. – Major, United States Marine Corps (1932-1967) *Plot: Section 3, Lot 2503 H-1, Grid Q-15*
Clifton Williams, a Marine Corps Major, graduated from the Navy Test Pilot School at Patuxent River, Maryland. He was test pilot for three years in the Carrier Suitability Branch of the Flight Test Division at Patuxent River. His work there included land based and shipboard tests of the F8E, TF8A, F8E (attack), and A4E and automatic carrier landing system. Of the 2,500 hours flying time accumulated, he has more than 2,100 hours in jet aircraft.

Williams was one of the third group of astronauts named by NASA in October 1963. He served as backup pilot for the Gemini 10 mission and worked in the areas of launch operations and crew safety. Major Williams died on 5 October 1967, near Tallahassee, Florida, in the crash of a T-38 jet.

The reference number is always on the reverse side.

Other related grave sites:

President John F 'Jack' Kennedy (1917-1963)
Plot: Section 45 Grid U-35

The second son of wealthy executive Joseph Patrick Kennedy and Rose Fitzgerald Kennedy, Jack Kennedy was born into a life of privilege but suffered from a variety of health ailments including whooping cough, appendicitis, and scarlet fever. His father taught his children to be competitive in every aspect of life, but stressed the role of public service. Jack was educated at private schools, and after his father was appointed US Ambassador to England, he studied at the London School of Economics. Following his return to Cape Cod, Massachusetts, he enrolled at Harvard University and majored in Political Science, with an emphasis on International Relations while participating on the school's football and swim teams. He later acquired experience while working as a secretary for his father in England.

After the United States entered World War II, he enlisted with the United States Navy. He rose to the rank of lieutenant and was commander of the PT-109. During a patrol off the Solomon Island, the boat was rammed into by a Japanese destroyer, resulting in the deaths of two crew members. The remaining personnel were thrown into the water. Kennedy swam for four hours to a nearby island while towing an injured sailor by his lifejacket. He then decided to swim out on his own in order to find help, and after several more hours in the water he reached another island where he met local anglers. He carved a message on a coconut and gave it to the anglers, who in turn forwarded it to the allies. Kennedy and his crew were soon rescued. The ordeal however, further damaged an already weakened back, which hindered Kennedy for the remainder of his life. During the war, Kennedy's older brother Joseph 'Joe' was killed in a plane crash. With Joe now gone, attention shifted to Jack to be groomed for a political career, and the ultimate goal.

In 1952, he was elected US Senator from Massachusetts, and in 1953 he married the beautiful Jacqueline Lee Bouvier, the daughter of a wealthy stockbroker, twelve years Kennedy's junior. Kennedy's popularity in the Senate grew to the point where he was strongly considered a running mate for 1956 Democratic Presidential Nominee Adlai Stevenson. He received a Pulitzer Prize for his book, 'Profiles in Courage' (1957).

Further health complications arose, including an ongoing battle with Addison's disease, which often sapped his strength. For his chronic back pain, he received regular injections of cortisone. In 1960, he launched his presidential campaign and promised a new chapter in America for a new generation born in the 20th Century. His Republican challenger was Vice President Richard Nixon. Their TV debate became historic. While Nixon looked uncomfortable and unsure, Kennedy came off as confident and focused. It is widely believed that the debate influenced many to be in favour of Kennedy. The 1960 Presidential Election was decided by the narrowest of margins (a little more than 100,000 votes) with Kennedy the winner.

During his inauguration speech, he included the now-famous line: *"Ask not what your country can do for you, ask what you can do for your country."*

Kennedy did not have to wait long for his first major challenge. In April 1961, tensions with Cuba's dictator Fidel Castro reached a breaking point. A plan was devised to liberate the country from Castro's ruling, orchestrated by the CIA. Kennedy approved the operation. The result was a total disaster and became what is known as *'The Bay of Pigs'*. He learned from his mistakes, and the following year during October 1962 another situation arose with Cuba. The Soviet Union had begun to establish missile sites on Cuba, aimed towards the United States. A strong possibility of nuclear war with the Soviets now existed. Over the following few days several messages would go back and forth between Kennedy and Nikita Khrushchev. Kennedy made a televised speech for implementing quarantine on Cuba until the bases were dismantled. The Soviets backed down and removed the sites. The event now known as *'The Cuban Missile Crisis'* showed Kennedy's ability to safely guide the US away from what could have been a world ending war. Our civilisation is currently living on borrowed time as a result.

The remainder of his time in the White House saw the Civil Rights Movement gain momentum. A situation that arose with Governor George Wallace of Alabama, who refused to let black students enter the University of Alabama, resulted in the need to use federal troops. Additionally, Kennedy established the Peace Corps.

After the Soviet Union launched its first satellite, then the first man in orbit in 1961, and along with their own atomic bomb tests, the US public became extremely concerned that they were vulnerable to an atomic war. To expand and improve US technology using non-aggressive means, Kennedy decided that a race to send astronauts to the moon would be a peaceful cover method of technological improvement. Project Apollo was born.

JFK along with Wernher von Braun at Cape Canaveral, Florida – now known as the Kennedy Space Center.

Fear of losing support in the South because of his policies towards Civil Rights led Kennedy to go on a trip to Dallas, Texas. During a motorcade in Downtown Dallas on 22 November 1963, accompanied by Mrs. Kennedy, Texas Governor John B. Connally and his wife Nellie, Kennedy was fatally shot. Over the next two days, his flag-draped casket would lie in state in the US Capitol Building. He was buried at Arlington National Cemetery on 25 November 1963. It is widely believed that Lee Harvey Oswald, the alleged assassin, did not act alone. Dallas nightclub owner Jack Ruby shot Oswald on live television two days after Kennedy's death. This added to the speculation of a conspiracy. In spite of this, Kennedy remains as one of the most popular presidents in US history and became the most quoted too.

John F Kennedy's grave is the first major site to be seen from the visitors' centre. Just walk out the back and follow the crowds uphill. His wife is buried beside him. His brother Robert Kennedy is buried nearby- to the left of the image.

Directions are simply not required.

George C Marshall (1880-1959)

Plot: Section 7, Grave 8198

George Marshall was a US Army General, US Secretary of State, US Secretary of Defense, and Nobel Peace Prize Winner. He served as the 50th Secretary of State from January 1947 until January 1949, and the 3rd Secretary of State from September 1950 until September 1951. He won the Nobel Peace Prize in 1953 for his development of the Marshall Plan – the European Recovery Program after the end of World War II.

Born into a Virginia family for several generations, he graduated from the Virginia Military Institute in 1901 and was commissioned a 2nd lieutenant in the US Army in February 1902. In his early career, he served as an infantry platoon leader and company commander in the Philippines during the Philippine–American War. From 1906 until 1910, he was a student and then an instructor at the US Army Command and Staff College at Fort Leavenworth, Kansas, during which time he was promoted to the rank of 1st Lieutenant. He was again promoted, this time to the rank of Captain in July 1916. When the US entered World War I in April 1917, he went to France as the director of training and planning for the 1st Infantry Division, and received a promotion to Major in August of that year. From 1918, he was posted to the Expeditionary Forces headquarters, where he worked with General John J. Pershing and was a key planner of American operations, including coordination of the Meuse-Argonne Offensive, which contributed to the final defeat of the German Army on the Western Front.

In 1927, as a Lieutenant Colonel, Marshall was appointed assistant commandant at Fort Benning, Georgia, and from June 1932 to June 1933 he was the commander at Fort Screven, Savannah Beach, Georgia (now Tybee Island) and was promoted to the rank of colonel after leaving that position. In October 1936, he was promoted to Brigadier General and commanded the Vancouver Barracks, Washington.

From 1936 until July 1938, he was assigned to the War Plans Division in Washington DC, and then reassigned as Deputy Chief of Staff. He attended a conference at the White House, at which President Franklin D. Roosevelt proposed a plan to provide aircraft to England in support of the war effort. With all other attendees voicing support of the plan, Marshall was the only person to voice his disagreement. Despite the common belief that he had ended his career, this action resulted in his being nominated by Roosevelt to be Army Chief of Staff. In September 1939 Marshall became a General, and would hold this post until the end of World War 2. He organised the largest military expansion in U.S. history, inheriting an outmoded, poorly equipped army of 189,000 men into a force of over 8,000,000 by 1942.

By mid-1943, after pressure from government and business leaders to preserve labour for industry and agriculture, Marshall had abandoned this plan in favour of a 90-division Army, using an individual replacement system sent via a circuitous process from training to divisions in combat. During World War 2, he was instrumental in preparing the US Army and Army Air Forces for the invasion of the European continent, and wrote the document that would become the central strategy for all Allied operations in Europe. He initially scheduled the invasion (codenamed Operation Overlord) for 1 April 1943, but met with strong opposition from British Prime Minister Winston Churchill, who convinced President Roosevelt to commit troops to Operation Husky for the invasion of Italy. It was assumed that Marshall would become the Supreme Commander of Operation Overlord, but Roosevelt instead selected General Dwight D. Eisenhower for the position. While Marshall enjoyed considerable success in working with Congress and President Roosevelt, he refused to lobby for the position, and Roosevelt did not want to lose his presence in the US.

On 16 December 1944, Marshall became the first US General to be promoted to a five-star rank, the newly created General of the Army. Throughout the remainder of World War 2, he coordinated Allied operations in Europe and the Pacific. In December 1945, President Harry Truman sent him to China to broker a coalition government between the Nationalist allies under Generalissimo Chiang Kai-shek and Communists under Mao Zedong. He had no leverage over the regime, but he threatened to withdraw American aid, which was essential to them at the time. Both sides rejected his proposals and the Chinese Civil War escalated, with the Communists eventually winning in 1949. Marshall returned to the US in January 1947. He was the appointed Secretary of State and became the representative for the State Department's ambitious plans to rebuild Europe.

On 5 June 1947, in a speech at Harvard University in Cambridge, Massachusetts, Marshall outlined the American proposal. The European Recovery Program became known as the Marshall Plan, and it helped Europe quickly rebuild and modernise its economy. The Soviet Union forbade its satellite countries to participate in the Plan. Marshall strongly opposed recognising the state of Israel, believing that if such a state was declared, that a war would break out in the Middle East (which it did in 1948, one day after Israel declared independence). In January 1949, he resigned from the State Department because of ill health, and the same month became chairperson of American Battle Monuments Commission. In September 1950, President Truman named him as Secretary of Defense. In this position, his main role was to restore confidence and rebuild the armed forces from the post-war state. Marshall served for one year and retired from public office in September 1951. He died 8 years later at the age of 78.

He was portrayed in a number of films, including 'Tora! Tora! Tora!' (1970) by actor Keith Anders, 'MacArthur' (1977) by actor Ward Costello, 'Saving Private Ryan' (1998) by actor Harve Presnell, and 'Pearl Harbor' (2001) by actor Scott Wilson.

On 15 March 1960, President Eisenhower announced that the space related complex within the borders of Redstone Arsenal would be named as the George C Marshall Space Flight Center. As part of the dedication ceremony, held on 8 September 1960, for the new Center, Mrs. Marshall and President Eisenhower unveiled a bust of General Marshall, which is still on display at the Space Center in Huntsville, Alabama.

Gary Francis Powers (1929-1977)
Plot: Section 11, Lot 685-2, Grid O/P-15.5
Born in Jenkins, Kentucky, raised in Pound, Virginia, Gary Powers graduated from Milligan College in eastern Tennessee. Upon

graduation in 1950, he enlisted into the US Air Force and was assigned to Turner Air Force Base, Georgia, flying F-84 Thunderjet Fighters. Shortly afterwards, he served in the Korean War, and in 1956, after promotion to Captain. He resigned from the USAF to join the Central Intelligence Agency, which was seeking Air Force pilots to join the U-2 program - a high-altitude reconnaissance plane, capable of flying 120,000 feet, well above the known limits of the then Soviet fighter aircraft and anti-aircraft missiles. CIA Pilots flew without identification in unmarked aircraft, and were issued a suicide pill if they were forced down.

Based at Incirlik Air Base, Turkey, Powers flew several flights over the Soviet Union, while the USSR was powerless to shoot down the planes. This changed by 1960, as in turn they also gained the ability to shoot down such highflying aircraft. On 1 May 1960, Powers took off from Peshawar, Pakistan, to fly across the Soviet Union and land at Boda, Norway, in a complete trip across the USSR. He was scheduled to return to Incirlik Air Base from Boda a few days later, returning via a different route. About four hours into the flight, his U-2 was severely damaged by the near miss of a Soviet surface-to-air missile near Sverdlovsk, Russia. Forced to bail out over Soviet territory, he was quickly captured before he had a chance to take his own life. President Eisenhower initially believed that Powers had died in the crash, and denied it was a spy mission until Soviet Premier Nikita Khrushchev held a press conference in which he produced a very much alive Powers, along with cameras and film from the plane.

After embarrassing President Eisenhower in front of the world press, Khrushchev then angrily cancelled the upcoming Paris Summit Meeting between the US and the USSR. Powers was tried for espionage, convicted and sentenced to ten years imprisonment. Twenty-one months after his capture, on 10 February 1962, he was exchanged for Soviet spy Colonel Rudolf Abel in Potsdam, Germany. Upon his return, the CIA, the USAF and Congress debriefed Powers, and all accepted that he did not reveal any classified information to the Soviets. In 1965, Powers was awarded the CIA Intelligence Star of Valour for his intelligence flights over the Soviet Union. However, the USAF broke their earlier promise to restore him to full officer status and to credit his time in the CIA and USAF (an offer they fulfilled to the other recruited pilots).

Powers returned to flying, working for Lockheed Aircraft as a test pilot from 1963 to 1970, before moving on to become a helicopter traffic reporter for Los Angeles TV station KNBC. He died in August 1977 when his helicopter crashed. Investigators found that it had a malfunctioning fuel gauge that they believed was the cause. Powers had written a book about his unique experience, *'Operation Overflight: A Memoir of the U-2 Incident,'* in 1970. Two movies were made; *The U2 Incident',* in 1970. Two movies were made: *The U2 Incident* with Lee Majors, and *A Bridge of Spies* with Tom Hanks.

Earl Warren (1891-1974)
Plot: Section 21, Lot S-32, Grid M-20.5

Earl Warren was a United States Supreme Court Chief Justice. As leader of the Supreme Court during a crucial period of US history, there is probably no other Chief Justice of the United States who evoked greater controversy in his time. Warren was born in Los Angeles, California, on 19 March 1891, the son of a Norwegian immigrant who worked for the Southern Pacific Railroad. Warren worked his way through college, receiving a Bachelor of Laws degree from the University of California in 1912. The young lawyer became a Deputy District Attorney in Alameda County, and was later elected District Attorney of the county in 1925. Prior to gaining the California governorship, Warren served as Attorney General from 1939-1943, gaining the image of an effective enemy of racketeers. However, Warren's role during WWII in orchestrating removal of persons of Japanese descent to internment camps was never forgiven by many.

After the war, Warren participated in Republican politics at the national level, serving as Thomas Dewey's vice-presidential running mate in 1948. Following the death of Chief Justice Fred M. Vinson on 8 September 1953, Eisenhower nominated Warren to the post of Chief Justice of the United States, out of gratitude for delivering the California vote in that presidential election.

In 1956, Warren wrote for a unanimous court in banning segregation in the nation's schools in the landmark ruling in Brown v. Board of Education. The 'Warren Court' proceeded to issue a stream of decisions striking down other aspects of segregation and broadening civil rights. As a result, the leader of that court soon became a target of angry conservatives.

In his memoirs, former president Richard Nixon reflected that the Warren Court had gone too far in *"attempting to remake American society according to their own social, political and ideological precepts."*

Warren is probably best known for the "Warren Commission" report formed by President Lyndon Johnson. The Chief Justice headed the effort to determine if the assassination of President John F. Kennedy was more than a one-man undertaking by Lee Harvey Oswald. The following year, the commission issued a report that concluded that no conspiracy existed.

The commission's investigation has been challenged through the years, and continues to this day to be considered as superficial by those advancing other theories. Earl Warren retired in 1969, and died at age 83 on 24 April 1993.

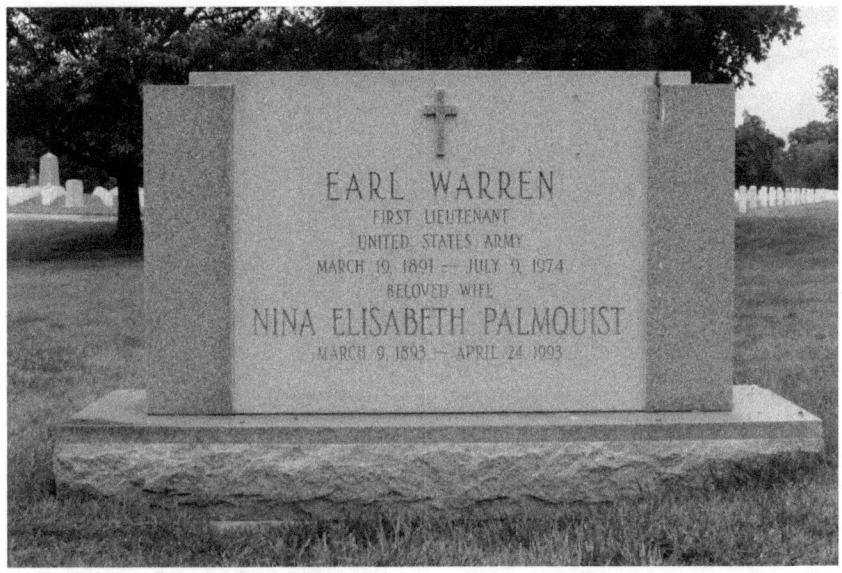

Shuttle Memorials

Not far from the grave of the Unknown Soldier are two Space Shuttle memorials for the Challenger and Columbia. Several of the astronauts on both flights are buried within 50ft beside and behind as mentioned previously. These memorials are easily found as they are well marked on all the Arlington guide maps.

Do not forget to observe the backs of these memorials or graves, where other engravings, quotes, images etc. are often portrayed.

The reverse side of both memorials

Also, lookout for the interesting wildlife that inhabit the cemetery.

Chapter 4   Einstein Memorial

Outside of the National Academy of Science

Einstein in Brief

Born on 14 March 1879, in the southern German city of Ulm, Albert Einstein grew up in a middle-class Jewish family in Munich. As a child, Einstein became fascinated by music, (he played the violin), mathematics and science. He dropped out of school in 1894 and moved to Switzerland, where he resumed his schooling and later gained admission to the Swiss Federal Polytechnic Institute in Zurich. In 1896, he renounced his German citizenship, and remained officially stateless before becoming a Swiss citizen in 1901.

Einstein developed the first of his ground-breaking Physics-related theories while working as a clerk in the Swiss Patent Office in Bern. After making his name with four scientific articles published in 1905, he went on to win worldwide fame for his General Theory of Relativity, and won the Nobel Prize in 1921 for his explanation of the phenomenon known as the photoelectric effect. An outspoken pacifist, who was publicly identified with the Zionist movement, Einstein emigrated from Germany to the United States when the Nazis took power before World War 2 fully broke out. He lived and worked in Princeton, New Jersey, for the remainder of his life.

In the late 1930s, Einstein's theories, including his equation, $E=mc2$, helped form the basis of the development of the atomic bomb and later atomic power. In 1939, at the urging of the Hungarian physicist Leo Szilard, Einstein wrote to President Roosevelt advising him to approve funding for the development of uranium before Germany could gain the upper hand. Einstein, who became a U.S. citizen in 1940 but retained his Swiss citizenship, was never asked to participate in the resulting Manhattan Project, as the U.S. government suspected his socialist and pacifist views. In 1952, Einstein declined an offer extended by David Ben-Gurion, Israel's Premier, to become president of Israel.

Throughout the last years of his life, Einstein continued his quest for a unified field theory. Though he published an article on the theory in Scientific American in 1950, it remained unfinished when he died of an aortic aneurysm five years later. In the decades following his death, Einstein's reputation and stature in the world of physics only grew, as physicists began to unravel the mystery of the so-called 'strong force' (the missing piece of his unified field theory) and space satellites further verified the principles of his cosmology.

Memorial

The Einstein Memorial features a 21ft high stylised bronze statue of physicist Albert Einstein, and sits amongst holly and elm trees on the Constitution Avenue side of the grounds of the National Academy of Sciences (NAS). This is not visible from the road, so if you wish to find it, first look for the path / sidewalk that leads to the NAS building, and follow the sign on the right hand side. Walk the trail about 20 yards or so, using the Lincoln Memorial as a guide to reach the NAS.

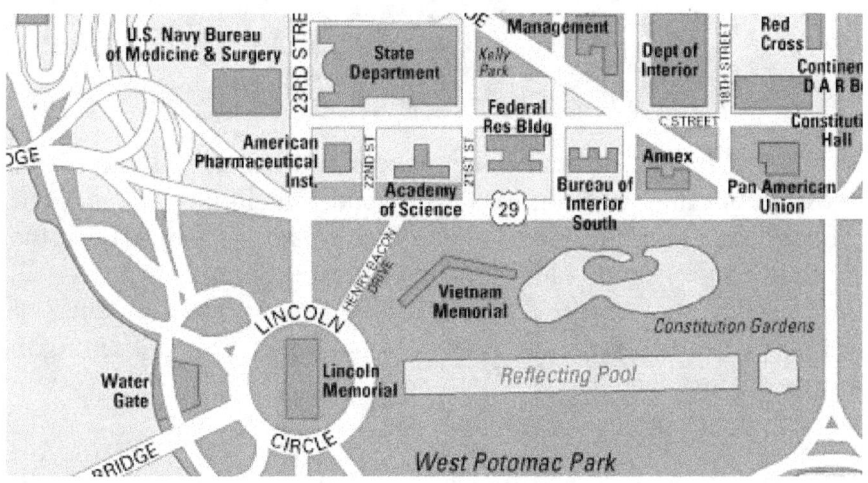

The memorial was unveiled on 22 April 1979, to commemorate the centenary of Einstein's birth. At Einstein's feet lies a granite base with studs that represent a star map of 2,700 stainless steel markers embedded into the marble to create a celestial map reflecting what was visible in the night sky on the day of the memorial's unveiling. This takes into account the light-travel distance to each object, just with the 'time' aspect removed. So, the position of each star on the map represents its *true* position in space on that date rather than the *apparent* position we see after the light has travelled perhaps for hundreds of years to reach us.

In his left hand, Einstein holds papers summarising three of his most important contributions to science, including his famous Theory of General Relativity. For all the conspiracy supporters, the words 'Theory' and 'Idea' have two very different meanings: an Idea is just that, but a Theory has to have ample evidence of support before calling it a Theory! Nobody has proved it wrong in over a century, and even the Internet or Global Positioning System (GPS) could not operate without his crucial formulas.

The main figure is three-times-life-size and consists of 3,500kg of bronze. The 'chewing gum' style by Robert Berks is not to everyone's taste, but it is certainly distinctive. 'Unfinished' is one of the kinder remarks. You may recognise the style – Berks also created the famous bust of President John F Kennedy at the Kennedy Center, also in DC, and others of Robert Kennedy at RFK stadium, and another of Abraham Lincoln at Ford's Theatre.

Chapter 5   Wernher Von Braun GraveSite

Ivy Hill Cemetery, 2823 King Street, Alexandria, VA 22302

Engineer and rocket expert Wernher von Braun was born in Wirsitz, Germany (now Wyrzysk, Poland) on 23 March 1912, to a wealthy family. After receiving a telescope from his mother at a young age, von Braun developed a passion for astronomy. In 1925, living with his family in Berlin, von Braun began reading Hermann Oberth's '*The Rocket into Interplanetary Space*', which spurred his desire to better understand science and maths, as the subjects related to space exploration. With his new dedication to his studies, von Braun became a top science student.

Von Braun enrolled at the Berlin Institute of Technology in the late 1920s, and graduated with a Bachelor's Degree in Mechanical Engineering in 1932. He then enrolled at the University of Berlin to study physics. While completing his graduate studies, von Braun conducted in-depth research on rocketry, for which he received a grant from the Ordnance Department of Germany. The grant financed von Braun's research at a research station not far from Berlin, next to the solid-fuel rocket facility of then-Captain Walter Dornberger, a department head for the Ordnance Department's armed forces. In 1934 von Braun obtained a Doctorate Degree in Physics from the University of Berlin. That same year, von Braun led a group that successfully launched two liquid-fuelled rockets that reached heights of more than 1.5 miles.

Moving to a new facility in the early 1940s in Peenemünde, a village in north-eastern Germany, von Braun worked with Dornberger and the rest of his crew, again successfully launching rockets as well as developing the supersonic anti-aircraft missile, Wasserfall, and the ballistic missile A-4. The A-4 became known as the 'V-2,' (Vengeance Weapon 2). Hitler was interested in using the V-2 for military purposes only, whereas von Braun wanted to launch the world's first satellite.

When von Braun refused to cooperate with Gestapo Chief Heinrich Himmler's attempted takeover of the V-2 project, and then criticised the Nazi regime via his dentist, he was imprisoned on espionage charges. Not long after, Hitler personally released von Braun, as he was so invaluable to the V-2 project. Despite never receiving approval from von Braun, German forces deployed the V-2 flying bomb against Britain, Holland, and Norway in 1944 - 1945.

In 1945, von Braun, as well as his brother, Magnus, and virtually the entire rocketry team, surrendered willingly to US troops. Signing a contract with the U.S. Army, von Braun was flown to Fort Bliss, Texas, then onto White Sands Missile Base in New Mexico, and eventually became Technical Director of the U.S. Army Guided Missile Project in Huntsville, Alabama in 1952.

Working alongside Dr William H. Pickering, former director of JPL, California, and Dr James A. van Allen, von Braun was an integral part of the team that successfully launched the first US artificial satellite, Explorer 1, on 31 January 1958. Leading the Army's Redstone Arsenal team, von Braun was responsible for the first stage Redstone Juno-I rocket that launched Explorer 1. In addition, under his direction, the Jupiter Intermediate Range Ballistic Missile and the Pershing missile were developed. During this period, von Braun becomes a legal U.S. citizen in 1955.

As director of NASA's Marshall Space Flight Center, from 1960 to 1970, von Braun developed the Saturn 1B and later the Saturn V rockets for the Apollo 8 Moon orbit mission in 1968 that led to the six Moon landings of 1969-72.

Because he was good looking, approached TV appearances with clever inventiveness, and possessed a unique history, von Braun was occasionally the butt of both humorous and serious verbal attacks regarding his association of former German scientists now working for the U.S. space program. Privately, it affected him deeply, but publicly he seemed to handle it with professionalism; an admirable quality.

In 1972, von Braun became Vice President at the aerospace company Fairchild Industries, Inc. A few years later, he founded the National Space Institute, aimed at gaining public support for space activities.

Von Braun died on 16 June 1977 of heart failure. In his last days, he did witness the first free flying tests of the Space Shuttle Enterprise at Edwards Air Force Base, California. Throughout his long career, von Braun received several U.S. honours, as well as awards from professional societies worldwide. He authored and co-authored various works on rocket science and physics. Today, von Braun is still considered one of the most important specialists in the field of rocketry and jet propulsion in the United States.

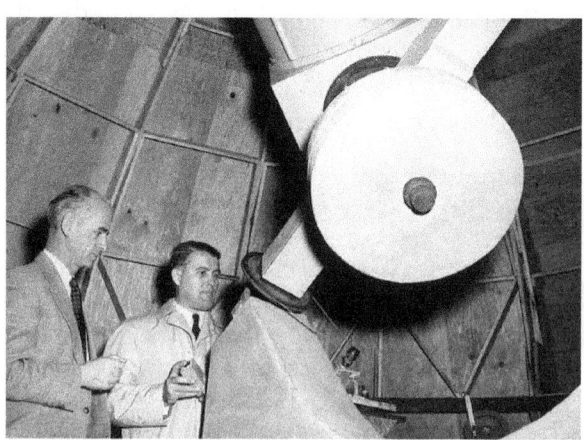

Von Braun at the Huntsville Astronomical Society's Observatory just east of the city in Alabama 1956.
I gave a lecture at the same site in 2006.

The Huntsville Astronomical Society was re-named the von Braun Astronomical Society. The society's Observatory includes a planetarium that is made from the top of a second stage spare fuel tank of a Saturn V rocket. Photos by the author.

Dr Wernher von Braun
Born: 23rd March 1912
Died: 16 June 1977
Section T, Lot: 29, Grave: 5

Ivy Hill Cemetery
2823 King Street
Alexandria, VA 22302
Phone: 703.549.7413

The Cemetery Grounds are open daily sunrise to sunset.
The Office is open Monday – Friday from 9 to 4, other times may
be available by appointment. Free onsite parking.

Walking from King Street Metro Station is about a 20-minute trek on sidewalks, and is mainly an uphill walk. From the Metro station, continue up King Street hill, keeping the Masonic Temple on your left. The Cemetery will be on your right-hand side once you have passed through the intersection with Janney's Lane. The entrance will be just before the beige and blue-trimmed two story 1850s farmhouse that serves as the Cemetery office. Taxis are available at the King Street Metro Station.

Several Alexandria DASH and Metrobus buses run past Ivy Hill Cemetery, and the stops are less than a block from the entrance of the cemetery. For more information about schedules and current fares, please refer to Dashbus, and for the Metrobus or subway transportation please refer to the Metro. DASH accepts regional SmarTrip Cards for fare payment. If you are on lower King Street, you can take the free King Street Trolley to the Metro station and proceed from there.

Chapter 6   US Naval Observatory

3450 Massachusetts Ave NW, Washington, DC 20392

www.usno.navy.mil

The U.S. Naval Observatory is one of the oldest scientific agencies in the US. It was established in 1830 as the Depot of Charts and Instruments. Its primary mission was to house the U.S. Navy's timekeeping instruments, charts and other navigational related equipment.

In 1844, as its mission evolved and expanded, the Depot became known as the U.S. Naval Observatory, and was located on a hill north of where the Lincoln Memorial now stands in Washington's Foggy Bottom district. Its role was modelled on the Greenwich Observatory in England. For almost 50 years, it remained at the Foggy Bottom location. Significant scientific studies conducted included speed-of-light measurements, solar eclipses, and transit of Venus expeditions that determined the accurate scale of the solar system. The astronomical and nautical almanacs commenced in 1855. In 1877, while working for the Naval Observatory, astronomer Asaph Hall discovered the two moons of Mars, Phobos and Deimos, as his wife burnt the dinner.

By the 1890s it was clear that the Observatory had to move out of the city. Unhealthy conditions in the Foggy Bottom neighbourhood had taken their toll. In 1893, after nearly 50 years at the site on the Potomac River, the U.S. Naval Observatory moved to its present location in the hilly terrain north of Georgetown. At that time, this rural site was well outside the city in the countryside above Georgetown. The move not only provided better astronomical observing conditions, but also an opportunity to rethink old scientific programs and propose new ones. Along with the new lines of research, such as daily monitoring of solar activity, the old functions of timekeeping and telescopic observations remained intact when the Observatory moved to the new site. The old Observatory in Foggy Bottom was declared a National Historic landmark in 1966, and is the current home of the Navy's Bureau of Medicine and Surgery.

Today, the U.S. Naval Observatory is still the authority in the areas of timekeeping and celestial observing, determining and distributing the timing and astronomical data required for accurate navigation and fundamental astronomy.

After nearly fifty years at the site on the Potomac River, hampered by fog and deteriorating buildings, in 1893 the U.S. Naval Observatory moved to its present location on Massachusetts Avenue in Northwest Washington, D.C. At that time, the site was well outside the city, separated from it by a deep valley. The renowned architect Richard Morris Hunt designed three of the buildings (the main building, the 26" dome and the transit circle buildings). Leon Dessez building, completed the Superintendent's residence, located to the north of the Observatory's main on his design.
Leon Dessez completed the Superintendent's residence, located to the north of the Observatory's main building, on Hunt's design.

In 1929, the Superintendent's residence became the home of the Chief of Naval Operations, and in 1974 Congress designated it as the Temporary Official Residence of the Vice President of the

United States. Today a high degree of security exists around the site as a result.

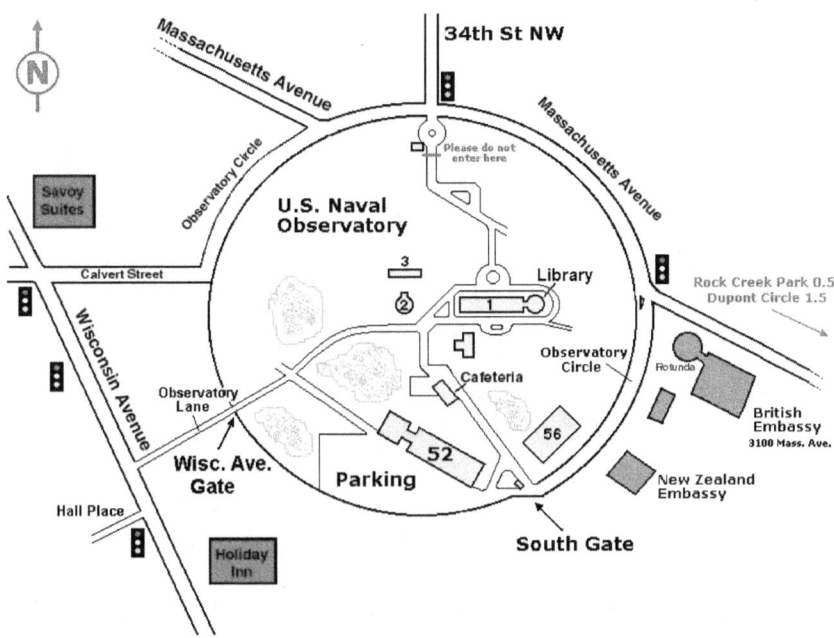

As an event that provided an opportunity to rethink old programs and to propose new ones, as well as in the provision of new facilities, the move to the new location was an important landmark in the history of the Observatory. Along with new programs, such as the daily monitoring of solar activity with a photoheliograph, (1899-1971), the old functions of timekeeping, meridian and equatorial observations remain intact. The move also provided the occasion for the Nautical Almanac Office, in Cambridge, Massachusetts since 1849 and located in Washington D.C. since 1866, to become officially a part of the Naval Observatory.

The challenge was now to achieve greater and greater accuracy in all areas of its mission, a quest that characterises much of the research at the U.S. Naval Observatory during the twentieth century. Greater accuracy required improved technology, and nowhere was this more evident than in the determination, maintenance, and dissemination of time. Beginning in 1934, the Observatory determined time with a photographic zenith tube (PZT), a specialised instrument that points straight upward toward the zenith and automatically photographs selected stars crossing the zenith. This gave a measure of the Greenwich Mean Time (now called Universal Time), the 'time of day' based on the rotation of the Earth. Improvements in clock technology, including the Shortt free-pendulum clock and quartz crystal clocks, soon proved conclusively that the Earth's rotation was not uniform, and a new uniform time-scale known as Ephemeris time came into use in 1956.

Defined by the orbital motion of the Earth about the Sun, in practice Ephemeris time was determined by observations of the Moon. This was undertaken by a dual rate Moon-camera, invented by William Markowitz at the Naval Observatory in 1951. In 1984, the family of time-scales known as dynamical time replaced Ephemeris time as the time based on the motion of celestial bodies according to the theory of gravitation, now considering relativistic effects. In the meantime, the development of atomic clocks replaced all previous timekeeping methods – the atomic time-scale being based on the vibration of the caesium atom.

In 1958, the Naval Observatory and Britain's National Physical Laboratory published the results of joint experiments that defined the relationship between Atomic time and Ephemeris time. (An interesting scientific and philosophical question is whether the relationship between atomic time and gravitational time remains constant). Since 1967, the international definition of the 'second' was based on these joint experiments. Atomic time is synchronised with universal time by the addition or subtraction of a leap second whenever necessary.

Time dissemination has been continuously improved. In 1904, a naval radio station transmitted the first radio time signals ever. They were derived from a U.S. Naval Observatory clock. This was the beginning of a system of radio time, a constantly improved and increasingly automated system that now spans the globe. The function of rating, repairing and disseminating chronometers and other nautical instruments, a major and especially critical effort during World War 2, was transferred from the Observatory to the Optical Section of the Norfolk Naval Shipyard in Portsmouth, Virginia in 1950.

The determination of the celestial coordinate system, against which the motions of all other celestial bodies must be measured, has been carried out at the Observatory by two transit circle telescopes (operated from 1894 to 1945) and the 9" telescope designed by William Harkness and mounted in 1899. The catalogues produced by these instruments are fundamental in the sense that in addition to selected stars, observations are also made of the Sun, Moon, planets, and asteroids.

The 6" transit circle has produced six fundamental star catalogues since 1924. This instrument has undergone many changes in technology to improve accuracy, from the method of reading its graduated circle by eye, to a travelling-wire micrometre, digital readouts of the measures, and then computerised data acquisition and telescope control. In 1956, a new 7" transit circle was installed to replace the 9".

In the quest for greater accuracy, the observatory has been constantly upgraded – with the 26" especially, as well as by the addition of new specialised telescopes. In 1935, G.W. Ritchey, a pioneer in telescope design who spent four years at the Naval Observatory on this project, completed a 40" Ritchey-Chretien reflecting telescope, one of the first of its kind. (The Hubble Space Telescope is of this same design.) This telescope was moved to the newly established Flagstaff station in Arizona in 1955, and was joined in 1963 by a powerful 61" astrometric reflector, designed and constructed under the direction of Director Kaj Strand. Again, the 61" is a pioneering design – the first, the biggest and the most accurate of its kind ever built. Together, the 40" and the 61" determine the relative positions, brightness, colours, and spectral types of stars, with electronic cameras, and by photography and photometry. The 61", which has a focal length of 50 feet, has, since its inception, carried out the world's largest program of determining stellar parallaxes; that is, accurate determinations of distances of nearby stars. For the first 20 years of its existence, it has concentrated on stars with magnitudes ranging from 12 to 18. For this program, 35 to 40 photographic plates were taken of each star, and the plates were then measured with an automatic measuring engine, which can determine star positions on photographs to better than one micron. This was the first distance measuring survey of its type.

The view of the Observatory from Arlington Bridge.
Photo by the author

The 61" reflector, used by James W. Christy, discovered the first satellite of Pluto (Charon) in 1978 – 101 years after Dr Asaph Hall found the moons of Mars from the same site. The discovery of Charon resulted in a precise determination of Pluto's mass, and which finally helped determine the path of the spacecraft New Horizons that visited the system in July 2015.

The 26" refractor was engaged throughout the 20th century in a program to observe natural satellites and double stars. Almost 30,000 visual measures of double stars were made in this program between 1961 and 1990. Since then, double stars have been observed with a technique known as speckle Interferometry. By taking very short exposures with a Charge-Coupled Device (CCD) camera, astronomers can actually use the blurring effect of Earth's atmosphere to their advantage to measure the separations and position angles of double star components. The technique ideally suited the telescope and was unaffected by the urban location. Several thousand stars are measured each year, and the database of such observations, added to the visual observations dating back over a century, provide for one of the finest double star catalogues in the world.

26" diameter lens Refracting telescope

The Nautical Almanac Office has fulfilled its essential function of predicting the positions of celestial bodies. Using the transit circle observations from the U.S. Naval Observatory and around the world, the Nautical Almanac Office has improved the theories of the orbital motions of solar system objects. These were used to construct tables for astronomers, navigators, and surveyors, printed for most of the 20th century as The Astronomical Almanac. For marine navigation, there is a separate publication, The Nautical Almanac.

For the use of air-navigators during World War 2, the Office designed and developed the American Air Almanac, first issued in 1941, and still issued today. Since its beginning, the Nautical Almanac Office has been in the forefront of the development and utilisation of computerized techniques in astronomy. This is necessary not only for the production of the Almanacs, and for providing astronomical data of various types for locations worldwide, but also for a wide range of research in celestial mechanics.

Above; the Observatory's Time Ball, based on the Greenwich, London design and purpose. Another such time-ball is found at Deal, Kent, UK with an associated museum.

The Observatory Today

The U.S. Naval Observatory continues to be the leading authority in the United States for astronomical and timing data, required for such purposes as navigation at sea, on land, and in space, as well as for civil affairs and legal matters. Its current Mission Statement, officially announced in 1984 by the Chief of Naval Operations, reads:

"To determine the positions and motions of celestial bodies, the motions of the Earth, and precise time. To provide the astronomical and timing data required by the Navy and other components of the Department of Defense for navigation, precise positioning, and command, control, and communications. To make these data available to other government agencies and to the general public. To conduct relevant research; and to perform such other functions or tasks as may be directed by higher authority."

Above; The Atomic Time Clock

The U.S. Naval Observatory, via its Directorates for Astrometry and Time, carries out its primary functions by making regular observations of the Sun, Moon, planets, selected stars, and other celestial bodies. Such measurements determine their positions and motions by:

a. Deriving precise time interval (frequency), both atomic and astronomical, and managing the distribution of precise time by means of timed navigation and communication transmissions, and ...

b. Deriving, publishing, and distributing the astronomical data required for accurate navigation, operational support, and fundamental positional astronomy.

The U.S. Naval Observatory conducts the research necessary to improve both the accuracy and the methods of determining and providing astronomical and timing data.

In addition to its Washington DC headquarters, the U.S. Naval Observatory maintains several field activities. The Time Service Alternate Master Clock Station at Schriever Air Force Base in Colorado serves as a backup to the Master Clock system in Washington, D.C. The Flagstaff Station provides a dark sky site at Flagstaff, Arizona, where the 61" reflector, the 40" reflector, a 24" reflector, and an 8" transit circle telescope, are located. An 8" astrograph, formerly stationed in Washington, has completed a complete CCD survey of the entire sky from the Cerro Tololo Observatory in Chile and Flagstaff, Arizona, which is now available as the USNO CCD Astrographic Catalogue. It is currently undergoing a complete renovation for the installation of a new 400-megapixel CCD/CMOS hybrid camera. This instrument will be deployed at Flagstaff and Cerro Tololo, and will be remotely operated.

Tips

The nearest tube station is probably Woodley Park-Zoo on the Red Metro line. Make sure you have a detailed road map to hand, as several roads in this area are very curvy and can easily disorientate you. This section is sometimes referred to as Embassy Row. Turn left from the British Embassy and walk around the circle until you reach the South Gate. Allow 25 minutes to walk from the Metro station to the gate.

My recommended option is Dupont Circle, also on the Red Line. It's a little further to walk, but just follow signs for Kalorama Heights in a straight line on Massachusetts Avenue as you come out of the station, then keep going; a much simpler route that doesn't even require a map. The visitor's entrance is again the South Gate off Massachusetts Avenue. Just turn left at the end of the road. A security check is required at the gate, so do take along some photo I.D. as this is a Federal site and often houses the US Vice President. Entrance is not always granted, especially if a government event is taking place. It is advisable to email in advance to avoid disappointment. Guided tours are only offered on a few days a year.

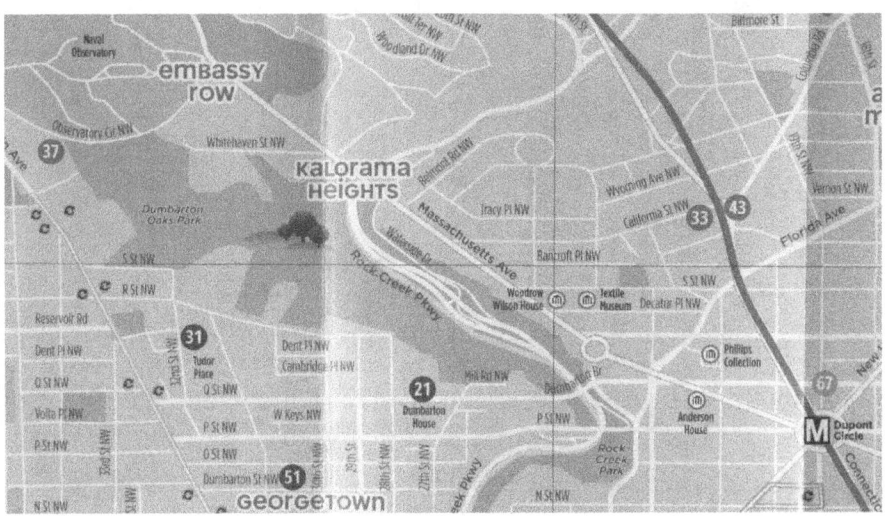

Chapter 7   National Air & Space Museum

600 Independence Ave SW, Washington, DC 20560
https://airandspace.si.edu

Every spaceflight enthusiast in the world must see this museum at least once. On a personal note, I first saw this in 1990. I arrived in DC early in the morning to beat the traffic, as I was completely new to the area. I parked at the Bureau of Engraving & Printing where the currency and stamps are designed and approved. I walked down the Mall on a light snow-covered sidewalk when it was still dark, and ran up the steps for a glimpse through the window at the exhibits. Within seconds I saw the original 1903 Wright Brothers plane, the Spirit of St Louis that Charles Lindbergh flew solo across the Atlantic, the Bell X1 that broke the sound barrier in 1947, John Glenn's Friendship 7 Mercury spacecraft, a Gemini spacecraft, and behind them the Apollo 11 spacecraft of the 1st Moon landing. It blew my mind, and I cried.

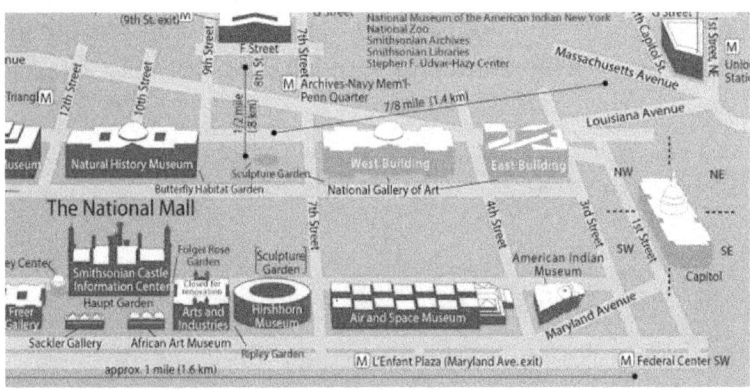

This place is easily found on the 12,000ft National Mall, so you cannot miss it. Face Capitol Hill, walk toward it, and just three blocks short is the Air & Space Museum on the right. Entry is free, but there is a security check as you enter, which consists of a bag search, that is all. Signs warn of not taking in tripods for cameras. With a little respect you can ignore this and just use your tripod discreetly. A monopod would be ideal.

History

In 1946, President Harry Truman signed a bill establishing the Smithsonian's National Air Museum to memorialise the development of aviation; to collect, preserve, and display aeronautical equipment, and to provide educational material for the study of aviation. The legislation did not provide for the construction of a new building, however, and the collection soon outgrew the Museum's exhibition space. Since there was no room left in the Arts and Industries Building or the *Tin Shed,* WW2 aircraft and other items, such as engines and missiles, were stored at an abandoned aircraft factory in Park Ridge, Illinois, a suburb of Chicago. The U.S. Navy had a similar collection in storage for the Smithsonian at Norfolk, Virginia.

In 1951, because of the Korean War, the Museum had to vacate the Park Ridge premises. In response to the immediate need for space, Paul Garber, the National Air Museum's first curator, located 21 acres in Silver Hill, Maryland, a suburb of Washington, D.C. With the addition of several prefabricated buildings, the site became the storage area for the National Air Museum. Garber had managed to save the collection.

Over the next 50 years, as aviation technology continued to advance, the collection expanded to include rocketry and spaceflight. It became clear that the Museum was entering a new phase. In 1966, President Lyndon Johnson signed a law that changed the name of the National Air Museum to the National Air and Space Museum. The Museum's collection on display expanded to include missiles, rockets and payloads, some of which were located outdoors near the Arts and Industries Building in an area that was known as 'Rocket Row'.

Funding to construct a new building was approved in 1971, and with the location determined, (it would be on the National Mall between Fourth and Seventh Streets S.W.), the Smithsonian Secretary, C. Dillon Ripley, hired former Apollo 11 astronaut Michael Collins as the National Air and Space Museum's director.

Collins would guide the Museum through its construction, hire a team of top-notch professionals, oversee the creation of first-rate exhibits, and launch the Museum's Center for Earth and Planetary Studies. This new division was devoted to active research in analysis of lunar and planetary spacecraft data, and the lead centre for Earth observations and photography from the Apollo-Soyuz Test Project.

Construction site of the National Air and Space Museum building on the Mall, 14 December 1972 – about the same time that the last Moon Landing was taking place – Apollo 17. A building on the near right became the NASA Headquarters for several years.

Ground-breaking of the new site took place on 20 November 1972, and in early 1975 the awesome task of filling the building with aircraft and spacecraft began. The goal of opening during America's bicentennial year was met, and the building was inaugurated with great fanfare on 1 July 1976.

The success of the Smithsonian's new National Air and Space Museum exceeded expectations. The five millionth visitor crossed the threshold only six months after opening day. Today, the

National Air and Space Museum is one of the most visited museums in the world.

The original Wright Brothers aircraft – The Flyer. Once held aloft in the main entrance, now has its own room stylised in 1903 architecture. Photo by the author.

A typical spaceflight nut admiring the Apollo 11 spacecraft that was occupied by Neil Armstrong, Buzz Aldrin & Mike Collins in July 1969; a rare occasion when it is not shrouded in Perspex. Photo by Amanda Bassett

Charles Lindbergh's Spirit of St Louis that flew him across the Atlantic solo, and an engineering mock-up of a Lunar Module that was used in training by the astronauts.

Gift Shop

Occupying three floors, this is well worth a visit for mugs and T-shirts. The top floor is almost dedicated to aviation and spaceflight books. You may even see this book on display. Many items are hard to find on-line, so get your cash and cards ready. Authors sometimes arrange a book-signing event with the store.

The middle and lower floors contain posters, patches, clothing, mugs, and the usual fridge magnets etc. There is one exhibit that is well worth seeing: the original Star Trek Enterprise model, as used in the TV first series with Captain Kirk and his loyal adventurous crew.

Einstein Planetarium

The Einstein Planetarium is worth a visit too, though do allow enough time for the rest of the museum. Visitors who are fans of Aviation as well as spaceflight should allow two days for this site to cover it thoroughly.

In 2005, I attended one particular planetarium show, and the lady presenter asked if anyone could pick out *'The Big Dipper'*. My hand went straight up, and she handed me the laser pointer. I showed the Big Dipper, then Polaris, Draco, Cassiopeia, Pegasus, Pisces, Cepheus, Hercules… at which point the laser was wrenched out of my hand. I apologised and explained that I was a Planetarium presenter from the UK. Everyone laughed. Surprisingly, I was invited to give a full talk on my next visit.

Never pass an opportunity, as you never know what it may lead to, and when it does, grab it with both hands – even if it is just a laser pointer.

Apollo 15: Dave Scott's Moon suit complete with Moon dust on full display. The other Moon walker on Apollo 15 was James Irwin; his suit is at Dulles Air & Space Center – next chapter; also free entry.

The Phoebe Waterman Hass Public Observatory

This observatory is found at the back of the museum, not far from the McDonalds end. The centrepiece of the Public Observatory is a 16" Boller & Chivens telescope, purchased in 1967 by Harvard College Observatory. It is named the Cook Memorial Telescope in memory of Chester Sheldon Cook, a long-time member of the Amateur Telescope Makers of Boston. The telescope was used by generations of students at the Oak Ridge Observatory in Harvard, Massachusetts.

With its closing in 2005, the telescope was loaned, and later fully donated, by Harvard to the Smithsonian Museum as the main instrument. The same site (as above) is now used to detect laser transmissions from sources of possible Extra-terrestrials.

It also houses several solar telescopes that allow the public to safely view the Sun in different types of light. With the white-light telescopes, you can see sunspots on the Sun's surface. The hydrogen-alpha (red light) and calcium-K (purple light) telescopes can reveal a variety of solar features in the Sun's atmosphere.

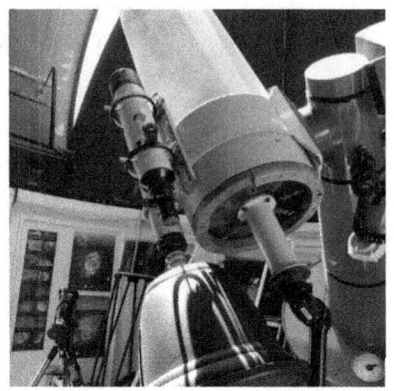

The observatory was named after Phoebe Waterman Haas. She received her doctorate in astronomy from the University of California, Berkeley, in 1913 — one of the first American women to earn such a degree.

Phoebe also studied at the historic Lick Observatory near San Jose, California, and she is believed to be the first woman astronomer to conduct her own telescopic research and not rely on the observations of others. It was named also after her in recognition of a $6 million donation from the Thomas W. Haas Foundation to establish the Museum's Public Observatory Program. It is the largest donation ever given to the Museum for science education programming. Phoebe Haas was the grandmother of the foundation's president, Thomas W. Haas.

Chapter 8   Dulles Air & Space Museum

(Steven F. Udvar-Hazy Center)
14390 Air and Space Museum Parkway
Chantilly, Virginia 20151
Phone: 703-572-4118
https://airandspace.si.edu/udvar-hazy-center

As the Smithsonian Air & Space Museum has been in operation since 1975, the number of possible exhibits had since expanded greatly. The end of the Apollo era, Concorde, the cold war, and from 2011, the end of the Shuttle era, meant that many large pieces of historic value had or will have no suitable home. A demand clearly arose for another site to display such items.

The Udvar-Hazy Center in Chantilly, Virginia, is built very close to Dulles International Airport. The sponsors change over the years, and so does the name, (tricky to remember too). Therefore, for the purpose of longer-term reference, I have linked it to Dulles, as it was when it first opened.

This has become the spin-off companion facility to the Smithsonian Museum on the National Mall in Washington, DC. It opened in 2003 incorporating two huge hangars — the Boeing Aviation Hangar and the James S. McDonnell Space Hangar. Between them they display thousands of extra aviation and space artefacts, including a Lockheed SR71 Blackbird, a Concorde, and the space shuttle Discovery (it replaced the Enterprise – now in New York Harbour on the USS Intrepid).

The Center also offers the Airbus IMAX Theatre and the Donald D Engen Observation Tower. From there it gives visitors a 360-degree bird's-eye view of Washington Dulles International Airport and the surrounding area. The Center is also home to the Mary Baker Engen Restoration Hanger. This is where preservation of the entire National Air and Space Museum's collections takes place. A glassed-in viewing area provides a view of various projects in progress. Researchers will also find the majority of the Archives' collections at the Udvar-Hazy Center reading room.

Entrance is completely free. As of 2017, the car parking charge is $15 for all day, but if you have visited before and just require a brief stop, then turn up after 4pm and the parking is free too. A full day is recommended to explore the complex properly. It does include a gift shop (for adults too for a change) and a canteen.

For Sci-Fi fans: on display is the UFO Mothership model from the movie Close Encounters of the Third Kind

The Enola Gay that dropped the first nuclear weapons in history during warfare.

The Original Space Shuttle Discovery. It launched the Hubble Space Telescope, and I was fortunate enough to witness the launch in 1990. The Columbia occupied the other shuttle pad at the same time; a rare occasion in itself. All images by the author.

Chapter 9 Natural History Museum

10th St. & Constitution Ave. NW, Washington, D.C. 20560
https://naturalhistory.si.edu/

The National Museum of Natural History is administered by the Smithsonian Institution, located on the National Mall just a couple of blocks from the Capitol Building on the left as you face it. It has free admission 364 days a year, (closed Christmas Day only). A bag search is required upon entry. It is the third most visited museum in the world, and the most visited natural history museum. Opened in 1910, the museum was one of the first Smithsonian buildings constructed to hold the national collections and associated research facilities. The main building has an overall area of 1,500,000 square feet, with 325,000 square feet of exhibition and public space, and houses over 1,000 employees.

The museum's collections contain over 126,000,000 specimens of plants, animals, fossils, mineral, rocks, meteorites, human remains, and human cultural artefacts. With 8,000,000 visitors in 2016, it is the most visited of all the Smithsonian museums, and is home to about 185 professional natural history scientists – currently the largest group of scientists dedicated to the study of natural and cultural history in the world.

Two massive iron meteorites from Arizona

The Space Stuff

Head for the mineral section and then onto the Meteorites. Samples come from all over the globe, and many are huge. Displays show clearly where the meteorites come from, how they formed and were examined after recovery. Amongst the largest is an iron meteorite from the Canyon Diablo Meteor Crater in Arizona. Other displays also include the Earth's interior, with many samples of lava from various layers in the Mantle.

In Antarctica, the ice slowly moves across land. Meteorites that have fallen onto the ice are dragged along and become caught up against mountain ranges. They tend to pile up over millions of years, and have become the largest concentration of meteorites anywhere on Earth. A display is dedicated to this region and shows a wide variety of such material.

Lunar Olivine Basalt collected by the Apollo 15 mission by either Jim Irwin or Dave Scott. It is around 3.2 billion years old. This is one of the largest pieces of Moon rock on public display in the world. I have seen around 60 out of the 200 or so samples. The largest I have ever seen was about 40cm across at the Lunar Receiving Laboratory in Houston, Texas – now sadly off limits to the public.

Chapter 10   NASA Headquarters

300 E St SW, Washington, DC 20546

www.hq.nasa.gov

Most people, when they say they have been to NASA, are normally referring to the Kennedy Space Center in Florida. However, the NASA HQ in Washington DC really is the heart of the NASA organisation that controls all the centres around the USA. The list below gives an indication of this building's power...

Ames Research Center
Armstrong Flight Research Center
Glenn Research Center
Goddard Space Flight Center
Goddard Institute of Space Studies
IV and V Facility
Jet Propulsion Laboratory
Johnson Space Center
Kennedy Space Center
Langley Research Center
Marshall Space Flight Center
Michoud Assembly Facility
NASA Engineering and Safety Center
NASA Safety Center
NASA Shared Services Center
Plum Brook Station
Stennis Space Center
Wallops Flight Facility
White Sands Test Facility... *phew!*

Unless you have a pre-arranged appointment with NASA for a specific reason, then you may want to give this a miss, unless you just want a selfie outside the building to prove you were there. You can get as far as the reception desk, and you may be lucky to see the gift shop open. It is small compared to the one in the Air & Space Museum, but does have some rarely seen items for sale. The opening hours are normally from 8am-4pm; do check the website for details. I have attended an interview at the NASA HQ regarding the promotion of my book, *'The Great Moon Landing Hoax – Or was it?'* (More details at the end of this book).

Directions from the Metro stations

Orange Line or Blue Line: To Federal Center SW – take the escalator to the 3rd Street exit, turn right, walk under the small overpass (1½ blocks) to the intersection of third and E Street. NASA HQ is on the right. All visitors must enter through the visitor's entrance, which is located on the West side of the NASA HQ building (four blocks behind the National Air & Space Museum on the Mall.

Yellow Line or Green Line: To L'Enfant Plaza - Follow signs to the 7th & D St exit. Once on D Street, turn right and walk half a block to 6th Street. Turn right and walk a block to E. Street. Turn left and walk 1½ blocks to 4th Street. NASA Headquarters is across the street on the right.

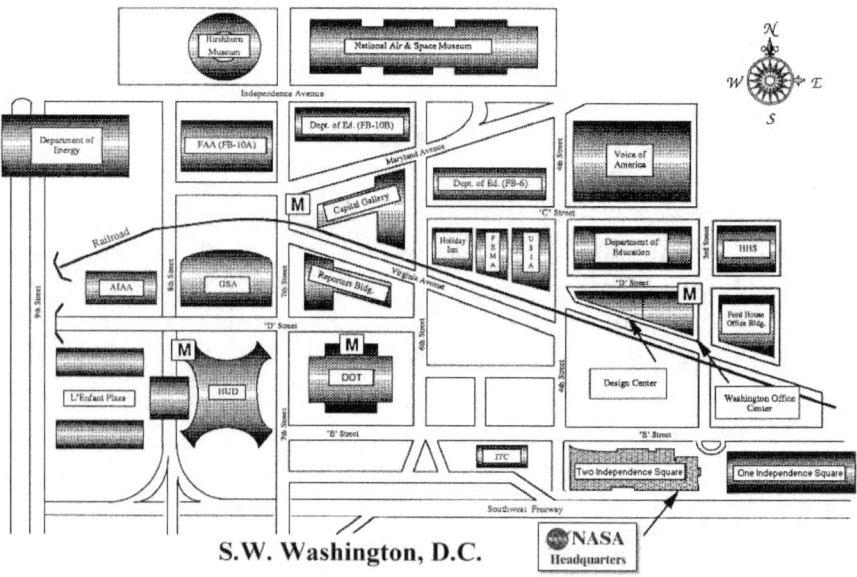

S.W. Washington, D.C.

Chapter 11 Capitol Building

The National Mall – cannot miss it!

Entry to the Capitol Building is free of charge but does require advanced booking online. This is not just for scheduling, but also for security clearance checks. A guided tour is offered of the ground floor and the basement. The final part is an inside view of the Dome. Cameras are allowed; just ensure yours is fully working with charged battery, as you may be asked to turn it on to prove it has not been hollowed out for another purpose. If you are not a US Citizen, have your passport ready.

The connection with space history is in three parts. First is a memorial to Rusty Swigert. He was a high-performance-jet pilot and later became an Astronaut. Then he was elected to enter the House of Representatives; hence the political connection.

Jack Swigert.
John (Jack) Swigert was born on 30 August 1931 in Denver, Colorado. He attended the Colorado University and earned a Batchelor of Science Degree in Mechanical Engineering. He became a combat pilot and flew fighter jets in Korea.

Jack was played by Kevin Bacon in Ron Howard's movie Apollo 13.

Swigert built on his degree toward aerospace science, and was accepted into the NASA Apollo Program. He felt at the time that this was going to be the peak of his career, and was assigned a trip

101

to the Moon – Apollo 13. This third Moon-landing attempt was aborted due to the rupture of an Oxygen tank that was a part of a Hydrogen fuel cell. It was later realised that it was a faulty module, and the associated paperwork had somehow gone astray. (It had in fact been placed in the Apollo 13 Command Service Module months beforehand). It was doomed to fail from that point, and *not* because of astronaut error during the mission, as shown in the movie *'Apollo 13'*.

After retiring from NASA, Swigert became staff director of the Committee of Science & Technology in the US House of Representatives. Then he was elected to lead the new Colorado Sixth Congressional District. However, he died on 27 December 1982, a few days before taking office.

JFK

It is claimed today that people know more about the assassination of John F. Kennedy than his life as a president. Just to fill in a blank spot in some people's knowledge, project Apollo, the thrust to the Moon, was largely kindled by JFK. He wanted a technical race in space of some form against the Soviet Union. This was to be a peaceful method of demonstrating US technological capabilities. Many of the engineers felt that a space station would be the answer, but Kennedy imagined that the Soviet Union would probably beat them to it just as they did with the other 'firsts' such as the first satellite, first living creature in space, first human in orbit, first probe to pass the Moon and so on.

Kennedy wanted to land a man on the Moon. It was a goal that was too advanced for the Soviet capabilities at the time, and so gave the US a chance to catch up and overtake. Kennedy even gave a time limit – the end of 1969. History records the rest After his assassination in 1963, Cape Canaveral, Florida, was re-named the John F. Kennedy Space Center.

JFK's body was flown from Dallas, Texas, on Air Force One, and landed just outside Washington DC. His casket lay on the floor of the Capitol Building on the 25 November 1963 on this special marker under the centre of the dome.

On the inside of the dome, a painting of the Wright brothers' aircraft is seen, and the 'eagle' next to it was represented on the Apollo 11 mission patch. This is the third space connection. Michael Collins traced the outline of the same pose from a National Geographic book. Photos by the author.

Chapter 12   NOAA Sites

The actual mailing address for job applications is in the heart of DC... National Oceanic and Atmospheric Administration, 1401 Constitution Avenue NW, Room 5128, Washington, DC 20230.

The following address has been the technical Headquarters of the *National Oceanographic & Atmospheric Association.* Satellite data is largely used for weather & climate change modelling. The building is actually home to The Airmen Memorial Building, and the NOAA rent several floors. Computer monitors showing the latest satellite data decorate the lobby.

5200 Auth Rd, Camp Springs, MD 20746, USA
Phone: +1 301-683-1314
Open Monday – Friday 9am - 5pm

As of 2012, most of the facilities and staff have moved to a new site – a 10-acre parcel within the University of Maryland Research Park Campus in Prince George's County, Maryland. The colocation with the University allows for close collaboration with the research community, and access to University resources.

The new complex is officially known as *National Oceanographic & Atmospheric Association Center for Weather and Climate Prediction, or* NCWCP... phew! The organisation is complex and has many branches around the nation.

Although there is nothing of importance there from a tourist's point of view, this building houses the administrators that co-ordinate the 12,000 staff.

The new technical centre is very close to the NASA Goddard Spaceflight Center, and they share many resources. 5830 University Research Ct, College Park, MD 20740.

The 268,000 square-foot building is home to more than 800 employees of NOAAs Center for Weather and Climate Prediction, who provide the nation with a broad range of environmental services – from predicting the hurricane season and El Niño/La Niña to forecasting ocean currents and large-scale rain and snowstorms. Billions of Earth observations from around the world flow through environmental models, developed and managed in the new building, that support the nation's weather forecasts.

Weather data is compiled from many sources from ground stations, weather balloons, amateur stations, and satellites. Scientists here also predict how hazardous materials move in the atmosphere, conduct air quality modelling, study climate variability, monitor and predict movement of volcanic ash, and research new ways to use satellite information to safeguard the environment. Scientists also monitor hurricane and tropical cyclones worldwide, analysing fire and smoke plumes from wildfires, which NOAA satellites track.

In the lobby, several monitors show the latest data from various satellites that are used for daily weather forecasts. As far as I am aware, no tours for the public are available. Do investigate, as the policy could change anytime.

The University of Maryland is a massive Campus. Feel free to explore for a great relaxing afternoon in the heart of this intellectual centre of space-related excellence.

Chapter 13　Hubble Science Institute Baltimore, Maryland.

3700 San Martin Drive, Baltimore, MD 21218

www.hubblesite.org

The science operations are co-ordinated and conducted by the Space Telescope Science Institute (STScI) in Baltimore. Hubble's science instruments act as the astronomer's electronic eyes. Light hits the primary mirror, is reflected onto the secondary mirror, and from there passes through a small opening in the primary mirror to the instruments. Once the telescope observes a celestial object, its onboard computers convert the image or data into long strings of numbers, which are transmitted down to Earth via the Tracking and Data Relay Satellite System (TDRSS). From TDRSS the data is sent to a dedicated ground station, and on to the Goddard Space Flight Center. NASA's Goddard Space Flight Center then forwards the data over land-links to the Space Telescope Science Institute (STScI) in Baltimore, where it is reconverted into data and images, analysed and stored. Observers can then study the data at STScI or at their home institutions.

The Hubble Science Institute: next to the John Hopkins University Hospital.

107

The telescope operates 24 hours a day. However, scheduling is tricky, as at certain times of the year a celestial object may not be visible because it appears too close to the Sun. Furthermore, Hubble is not in constant communication with ground stations, and technicians must plan each observation down to the second, instructing the telescope what to do and when, using detailed computer-coded instructions transmitted to and stored inside the telescope's computer.

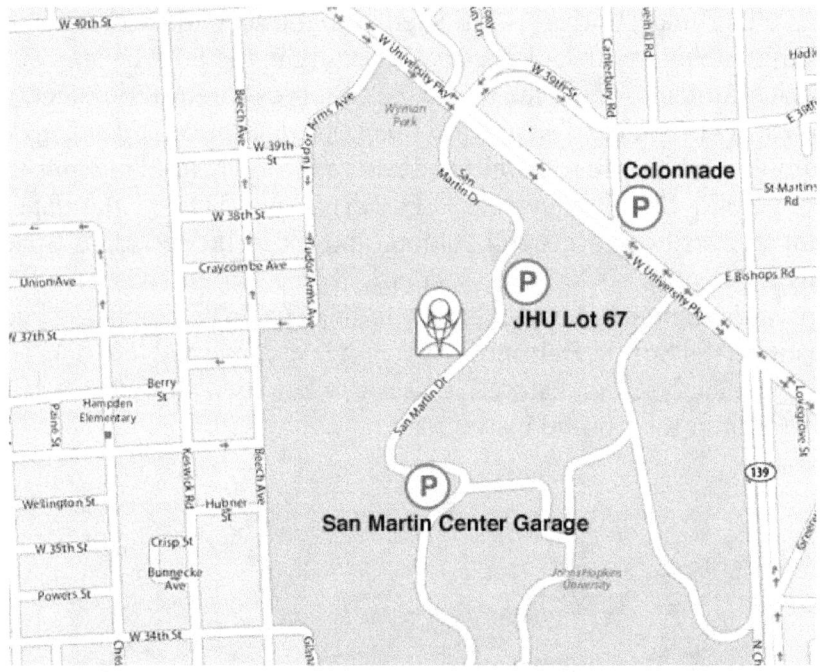

Muller Building (Main)
3700 San Martin Drive, Baltimore, MD 21218
Telephone: 410-338-4700

Although Hubble operates around the clock, not all of its time is spent observing. Each orbit lasts about 97 minutes, with time allocated for housekeeping functions and for observations. These include: turning the telescope to acquire a new target, or avoiding the Sun or Moon; switching communications antennae and data transmission modes; receiving command loads and down-linking data, as well as calibrating and similar activities.

When STScI completes its master observing plan, the schedule is forwarded to Goddard's Space Telescope Operations Control Center, (STOCC), where the science and housekeeping plans are merged into a detailed operations schedule. Each event is translated into a series of commands to be sent to the onboard computers. Computer loads are up-linked several times a day to keep the telescope operating efficiently.

If an astronomer wishes to be present during the observation, there is a console here where monitors display images or other data as the live observations occur. Some limited real-time commands for target acquisition or filter changes can be given from these stations, if the observation program has been set up to allow for it, but spontaneous last-minute control is not possible.

NASA image

Tourists

For the average tourist, this may not be the most exciting place to visit. Anyone can just walk in off the street and see a few monitors in the lobby showing live and historic data from the telescope. One can ask for leaflets and posters on the site and the mission, (hint: tell them you are in the education business), but pre-arranged appointments are required to go any further. At least you can say you have been there and can prove so with a quick selfie picture.

Chapter 14   Goddard Spaceflight Center Greenbelt, Maryland

8800 Greenbelt Road, Greenbelt, MD 20771

www.nasa.gov/goddard

The Visitor Center is located off ICESat Road. Once on ICESat Road, turn left into the Visitor Center prior to the security checkpoint. Bags are searched before entry.

Opening times;
September through June
Tuesday - Friday: 10am - 3pm
Saturday, Sunday: noon - 4pm

July through August
Tuesday - Friday: 10am - 5pm
Saturday, Sunday: noon - 4pm

Closed almost every Monday.

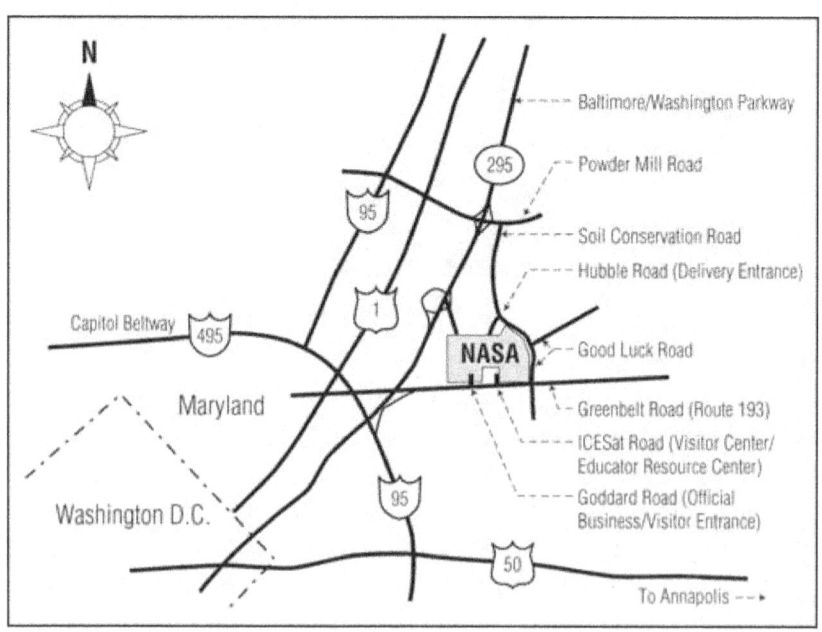

The Gift Shop is located next to the Visitor Center and has a wide array of NASA-themed items, including clothing, space-related toys and games, model rockets, a large selection of NASA pins and mission patches (the cheapest I have ever seen) and astronaut ice cream. Call for store hours on 301-286-6476. It is open for virtually every hour the Visitor Center is open; depending upon availability of staff.

Rocket Garden

The NASA Goddard Visitor Center features a full-size rocket garden located outside, which features many types of rockets, mock-ups, and old flight hardware. Every piece in the collection is a real NASA artefact that offers excellent photo opportunities. The Rocket Garden is also a great way to view the actual Goddard Space Flight Center, as the garden offers a sweeping panoramic view of the main campus.

Solarium Exhibit

From 11 Feb 2015, an innovative new piece of video art puts you directly in the heart of a mesmerising show. The display taps into a vast reservoir of imagery from NASA's Solar Dynamics Observatory (SDO). This data streams in hourly at Goddard.

SDO observes ultra-violet light to track how material travels through the solar atmosphere. SDO takes a picture almost once a second – no other solar observatory has ever collected data on the entire Sun with such regularity. Each image has eight times as much resolution as HD TV.

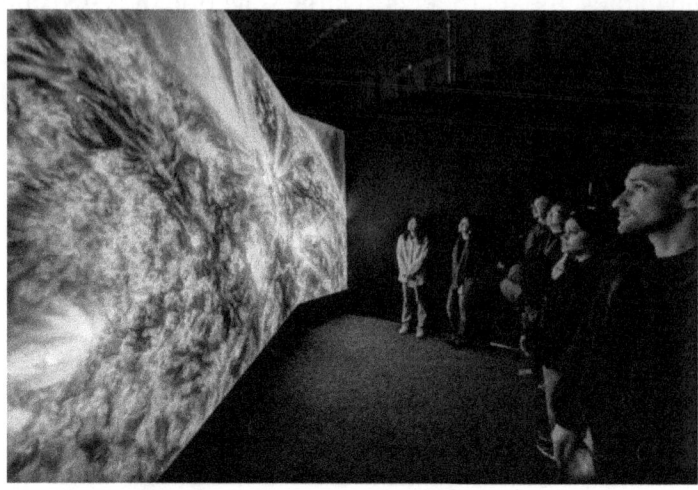

Hubble

Hubble's flight controllers work at the Space Telescope Operations Control Center (STOCC). This is based at the NASA Goddard Space Flight Center in Greenbelt, Maryland. From the STOCC, these highly experienced engineers and technicians 'drive' the telescope 24 hours a day, 7 days a week.

Hubble has four flight control teams, each consisting of three operators backed by dozens of engineers and scientists. The teams work in seamless shifts to keep the telescope productive and running smoothly. Using Hubble's Control Center System, they send more than 100,000 separate instructions to Hubble each week.

The STOCC is 'Hubble Mission Control' – the focal point of all Hubble Space Telescope operations. Controllers carry out routine operations in one section, while another section supports past servicing missions, including testing and simulations. In an adjacent section, engineers perform in-depth subsystem analysis, simulated tests, and integrate new databases, testing new ground and onboard Hubble software. Shuttle crews supplied new hardware, but as of 2011 no vehicle is capable of carrying out any further servicing. Once the Hubble breaks down, that will be the end of its amazing record-breaking mission.

The site is just accessible on the Metro: use either the Green or the Yellow lines. A taxi will be required to reach the Visitor Center entrance.

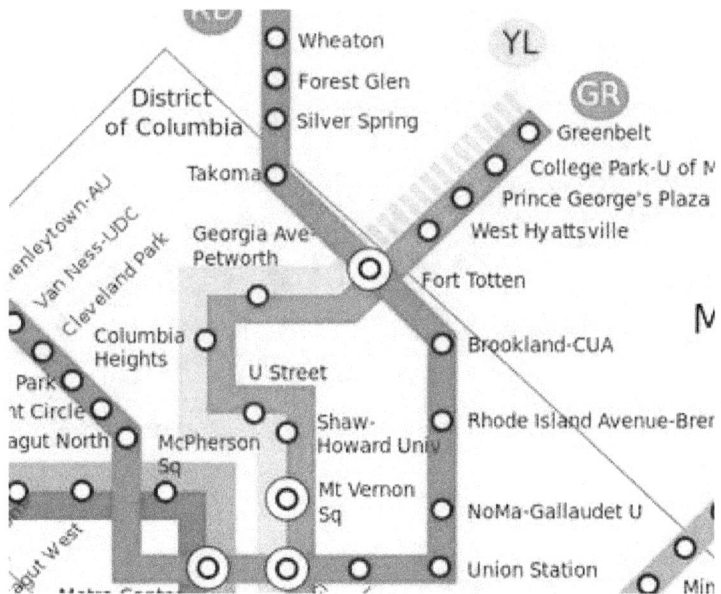

Chapter 15   Dining out

Walking around Washington DC requires a lot of physical energy. Food sources are not obvious in this town – apart from three burger joints. One is in front of the Lincoln Memorial, and the other two are along the Mall as you walk toward Capitol Hill. They do provide salad options, soft drinks and sandwiches. Beware of the pigeons begging for scraps – they can become a real nuisance and ruin your day. (N.B. From hard-earned experience, they do not care where they poop).

If you want something more civilised, then other food outlets are not so easy to find. The next option is probably the McDonald's restaurant in the Smithsonian Air & Space Museum. It is the most expensive I have ever seen though, and is often a chaotic experience at that particular branch. There may also be another bag search if you re-enter the Museum, as they ban uneaten food.

Union Station

Another food court is in Union Station, a short walk behind and to the north of the Capitol Building, just two blocks away. It has several fast-food outlets and healthier options on offer. Gift shops are there too. You may be lucky and time your visit with a lecture. Astronauts and pilots have given talks here for free in a very inter-active atmosphere with the audience.

Ronald Reagan Building

1300 Pennsylvania Ave NW, Washington, D.C.

Facing toward Capitol Hill with the Washington Monument behind you by a couple of blocks, on the left at 14th Street, turn down it to the Ronald Reagan Building, next to the Woodrow Wilson Center. The Federal Triangle is the nearest Metro Station. Choose any entrance, go through security, and find signs for the Food Court. Pass downstairs and discover an array of around 20 vendors including Subway and many foreign food outlets, (and no pigeons to pester you).

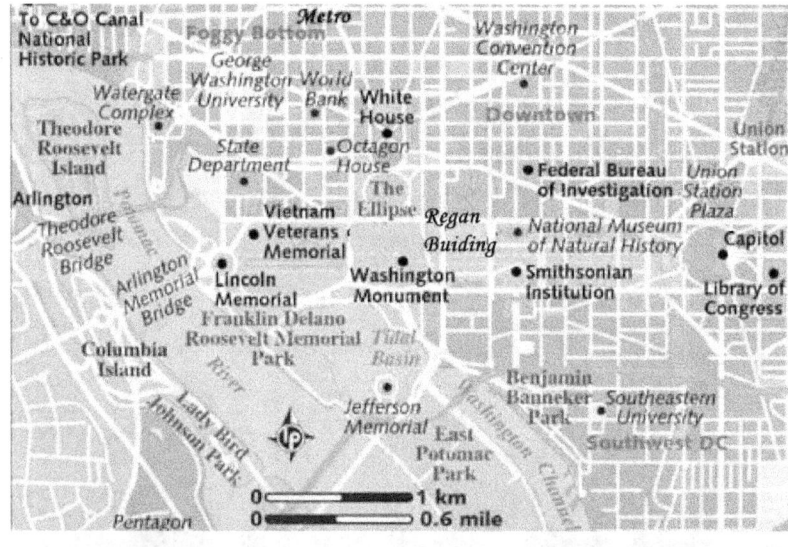

Crystal City

Over the years, my wife and I have refined our visits in detail to make each experience as hassle-free as possible. Just three stops from the Pentagon is an underground complex known as Crystal City. It has a dedicated Metro station on the Blue & Yellow lines. One of the best deals for a hotel anywhere around DC was two blocks from this station: the Americana. The price included all-you-can-eat breakfast, and ran a shuttle service to and from Dulles Airport if required. Sadly in 2023, it was demolished. Finding a hotel within walking distance to a Metro Station would be strongly advisable.

Walk underground to the east for several blocks until you reach the food court. It includes two 'Subways' and plenty of further choices.

23rd Street is home to a strip of restaurants for all tastes and budgets. Our favourite became Bob & Edith's Diner, offering many healthy options, free Wi-Fi, which is great if you are keeping costs down. This is a recent addition to the area. 23rd Street includes banks, liquor outlets and a 7 Eleven general store.

Around the next block is S Eads Street, and more restaurants are found along there too. This is colourfully lit at night, complete with a mini waterfall, live entertainment on occasions, and outside seating.

23rd Street – perfect for any kind of dining and suits all budgets.

Chapter 16   Links for Reference

All the internet links in this book are listed here for reference:

Publisher's main website:
www.astronomyroadshow.com

Renting Bicycles: www.capitalbikeshare.com

Tour Busses: www.trolleytours.com

Arlington Cemetery: www.arlingtoncemetery.mil

US Naval Observatory: www.usno.navy.mil

Hubble Science Institute: www.hubblesite.org

Goddard Spaceflight Center: www.nasa.gov/goddard

Dulles Air & Space Museum:
www.airandspace.si.edu/udvar-hazy-center

Natural History Museum: https://naturalhistory.si.edu/

NASA Headquarters: www.hq.nasa.gov

NOAA main website: www.ncdc.noaa.gov/

National Cherry Blossom Festival:
www.nationalcherryblossomfestival.org

Chapter 17   Other books by the Author

New books to be published in the coming years.

www.outerspacebooks.com

For signed copies, updated listings and direct website links. Most have a supporting website. The following are amongst those available…

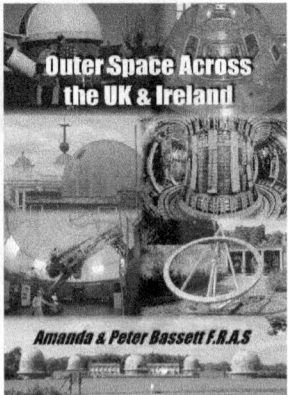

**There are B&W, Colour and e-book versions.
All book sales support four charities:**

Cancer Research UK
Kent Air Ambulance
Smile Malawi Orphanage in Africa
British Hedgehog Preservation Society

*Amanda & Peter Bassett trying to look cool
with the Capitol Building behind: 2009.*